TESTING COMPLEX AND EMBEDDED SYSTEMS

Kim H. Pries
Jon M. Quigley

CRC Press
Taylor & Francis Group
Boca Raton London New York

CRC Press is an imprint of the
Taylor & Francis Group, an **informa** business
AN AUERBACH BOOK

CRC Press
Taylor & Francis Group
6000 Broken Sound Parkway NW, Suite 300
Boca Raton, FL 33487-2742

© 2011 by Taylor and Francis Group, LLC
CRC Press is an imprint of Taylor & Francis Group, an Informa business

No claim to original U.S. Government works

Printed in the United States of America on acid-free paper
10 9 8 7 6 5 4 3 2 1

International Standard Book Number: 978-1-4398-2140-4 (Hardback)

Library of Congress Cataloging-in-Publication Data

Pries, Kim H., 1955-
    Testing complex and embedded systems / Kim H. Pries, Jon M. Quigley.
        p. cm.
    Summary: "Using combinatorial approaches, this book aims to motivate testers and testing organizations to perform meaningful testing. The text details planning activities prior to testing, how to scope the work, and how to achieve a successful conclusion. Rather than presenting the entire continuum of testing for a particular product or design attribute, this volume focuses on boundary conditions. The authors provide various techniques that can be used to streamline testing and help identify problems before they occur, including turbocharge testing methods from Six Sigma. Coverage includes testing, simulation, and emulation"-- Provided by publisher.
    Includes bibliographical references and index.
    ISBN 978-1-4398-2140-4 (hardback)
    1. Embedded computer systems--Testing. I. Quigley, Jon M. II. Title.

TK7895.E42P738 2010
004.16--dc22   2010043713

Visit the Taylor & Francis Web site at
http://www.taylorandfrancis.com

and the CRC Press Web site at
http://www.crcpress.com

TESTING COMPLEX AND EMBEDDED SYSTEMS

BOOKS ON SOFTWARE AND SYSTEMS DEVELOPMENT AND ENGINEERING FROM AUERBACH PUBLICATIONS AND CRC PRESS

Design and Safety Assessment
of Critical Systems
Marco Bozzano and
Adolfo Villafiorita
978-1-4398-0331-8

Implementing and Developing
Cloud Computing Applications
David E. Y. Sarna
978-1-4398-3082-6

Secure Java: For Web Application
Development
Abhay Bhargav and B. V. Kumar
978-1-4398-2351-4

Scrum Project Management
Kim H. Pries and Jon M. Quigley
978-1-4398-2515-0

Engineering Mega-Systems:
The Challenge of Systems
Engineering in the
Information Age
Renee Stevens
978-1-4200-7666-0

Certified Function Point
Specialist Examination Guide
David Garmus, Janet Russac, and
Royce Edwards
978-1-4200-7637-0

Enterprise Systems Engineering:
Advances in the Theory and
Practice
George Rebovich, Jr.,
and Brian E. White
978-1-4200-7329-4

Process-Centric Architecture for
Enterprise Software Systems
Parameswaran Seshan
978-1-4398-1628-8

Secure and Resilient Software
Development
Mark S. Merkow and
Lakshmikanth Raghavan
978-1-4398-2696-6

Real Life Applications of
Soft Computing
Anupam Shukla, Ritu Tiwari,
and Rahul Kala
978-1-4398-2287-6

Product Release Planning:
Methods, Tools and Applications
Guenther Ruhe
978-0-84932620-2

Process Improvement and
CMMI® for Systems and Software
Ron S. Kenett and Emanuel Baker
978-14200-6050-8

Applied Software Product
Line Engineering
Kyo C. Kang, Vijayan Sugumaran,
and Sooyong Park
978-1-42006841-2

CAD and GIS Integration
Hassan A. Karimi and Burcu Akinci
978-1-4200-6805-4

Applied Software Product-Line
Engineering
Kyo C. Kang, Vijayan Sugumaran,
and Sooyong Park, eds.
978-1-4200-6841-2

Enterprise-Scale Agile Software
Development
James Schiel
978-1-4398-0321-9

Handbook of Enterprise Integration
Mostafa Hashem Sherif, ed.
978-1-4200-7821-3

Architecture and Principles of
Systems Engineering
Charles Dickerson, Dimitri N. Mavris,
Paul R. Garvey, and Brian E. White
978-1-4200-7253-2

Theory of Science and Technology
Transfer and Applications
Sifeng Liu, Zhigeng Fang,
Hongxing Shi, and Benhai Guo
978-1-4200-8741-3

The SIM Guide to Enterprise
Architecture
Leon Kappelman
978-1-4398-1113-9

Getting Design Right:
A Systems Approach
Peter L. Jackson
978-1-4398-1115-3

Software Testing as a Service
Ashfaque Ahmed
978-1-4200-9956-0

Grey Game Theory and Its
Applications in Economic
Decision-Making
Zhigeng Fang, Sifeng Liu,
Hongxing Shi, and Yi LinYi Lin
978-1-4200-8739-0

Quality Assurance of
Agent-Based and
Self-Managed Systems
Reiner Dumke, Steffen Mencke,
and Cornelius Wille
978-1-4398-1266-2

Modeling Software Behavior:
A Craftsman's Approach
Paul C. Jorgensen
978-1-4200-8075-9

Design and Implementation of
Data Mining Tools
Bhavani Thuraisingham, Latifur Khan,
Mamoun Awad, and Lei Wang
978-1-4200-4590-1

Model-Oriented Systems
Engineering Science:
A Unifying Framework for
Traditional and Complex Systems
Duane W. Hybertson
978-1-4200-7251-8

Requirements Engineering for
Software and Systems
Phillip A. Laplante
978-1-4200-6467-4

TESTING COMPLEX AND EMBEDDED SYSTEMS

Kim H. Pries
Jon M. Quigley

CRC Press
Taylor & Francis Group
Boca Raton   London   New York

CRC Press is an imprint of the
Taylor & Francis Group, an **informa** business

AN AUERBACH BOOK

CRC Press
Taylor & Francis Group
6000 Broken Sound Parkway NW, Suite 300
Boca Raton, FL 33487-2742

© 2011 by Taylor and Francis Group, LLC
CRC Press is an imprint of Taylor & Francis Group, an Informa business

No claim to original U.S. Government works

Printed in the United States of America on acid-free paper
10 9 8 7 6 5 4 3 2 1

International Standard Book Number: 978-1-4398-2140-4 (Hardback)

Library of Congress Cataloging-in-Publication Data

Pries, Kim H., 1955-
    Testing complex and embedded systems / Kim H. Pries, Jon M. Quigley.
       p. cm.
    Summary: "Using combinatorial approaches, this book aims to motivate testers and testing organizations to perform meaningful testing. The text details planning activities prior to testing, how to scope the work, and how to achieve a successful conclusion. Rather than presenting the entire continuum of testing for a particular product or design attribute, this volume focuses on boundary conditions. The authors provide various techniques that can be used to streamline testing and help identify problems before they occur, including turbocharge testing methods from Six Sigma. Coverage includes testing, simulation, and emulation"-- Provided by publisher.
    Includes bibliographical references and index.
    ISBN 978-1-4398-2140-4 (hardback)
    1. Embedded computer systems--Testing. I. Quigley, Jon M. II. Title.

TK7895.E42P738 2010
004.16--dc22
                                                  2010043713

Visit the Taylor & Francis Web site at
http://www.taylorandfrancis.com

and the CRC Press Web site at
http://www.crcpress.com

Contents

List of Figures

List of Tables

Preface

This book is a collection of our thoughts and learning over several decades of test practice. Both of us have managed test groups in the automotive world, which includes the following disciplines:

- Software engineering
- Mechanical engineering
- Electrical engineering
- Metallurgical and materials engineering
- Procurement/purchasing
- Quality engineering
- Digital electronics
- Industrial engineering
- Manufacturing engineering

In part of the book, we compare contradictory perspectives to see if we can shed some light on beneficial considerations during the creation of our test suites and execution of our test plans. We know that system-level testing is usually the last-executed kind of testing and that this collection of final actions is often too late and too little, with much of the work accomplished in a rush.

Acknowledgment

We would like to thank John Wyzalek, the acquisitions editor for the Taylor & Francis Group of publishing companies; he has been a great help with development encouragement and promotion of all books we have written for them. We would also like to thank the rest of the people at the Taylor & Francis Group who have contributed throughout the process to produce this book.

I (Jon) would like to thank all the members of the Electrical and Electronic Verification and Test Group at Volvo 3P in Greensboro, North Carolina. We have had many discussions on how to improve product testing given the numerous challenges; there is no doubt that these discussions have influenced this book's content.

Last but not the least, I would like to thank my family. My wonderful wife Nancy, and my son Jackson, who is the best boy a daddy can have.

I (Kim) would like to thank my wife Janise Pries for reviewing this work—regardless, all mistakes belong to Jon and me. She is the love of my life and the reason I work such long hours to make a change in the world.

We used open source tools as much as possible to help develop this book; examples are

- winmerge (compare)
- mikTeX (LaTeX compiler)
- TeXnicCenter (TeX-specific editor)
- zscreen (screen capture)
- vim (programmer's editor)

About the Authors

Kim H. ("Vajramanas") Pries has four college degrees: a B.A. in history from the University of Texas at El Paso (UTEP), a B.S. in metallurgical engineering from UTEP, an M.S. in metallurgical engineering from UTEP, and an M.S. in metallurgical engineering and materials science from Carnegie Mellon University. In addition to these degrees, he has the following certifications:

- APICS
  - Certified Production and Inventory Manager (CPIM)
- American Society for Quality (ASQ)
  - Certified Reliability Engineer (CRE)
  - Certified Quality Engineer (CQE)
  - Certified Software Quality Engineer (CSQE)
  - Certified Six Sigma Black Belt (CSSBB)
  - Certified Manager of Quality/Operational Excellence (CMQ/OE)
  - Certified Quality Auditor (CQA)

Pries worked as a computer systems manager ("IT"), a software engineer for an electrical utility, a scientific programmer on defense contracts; and for Stoneridge, Incorporated (SRI), he has worked for 15 years as:

- Software manager
- Engineering services manager
- Reliability section manager
- Product integrity and reliability director

In addition to his other responsibilities, Pries provides Six Sigma training for both UTEP and SRI and cost reduction initatives for SRI. Additionally, in concert with Quigley, Pries is the co-founder and principal with Value Transformation, LLC, a training, testing, and product development consultancy. He is also a lay monk in the Soto tradition of Zen Buddhism and functions as an Ino for the Zen Center of Las Cruces while studying for the Soto Zen priesthood.

Pries' first book was *Six Sigma for the Next Millennium: A CSSBB Guidebook*, revised as *Six Sigma for the New Millennium: A CSSBB Guidebook*, second edition, from ASQ's Quality Press. For Taylor & Francis, Pries has worked with Jon Quigley to write *Project Management of Complex and Embedded Systems, Scrum Project Management*, as well as this book. With Quigley, he has written well over 30 magazine articles. He has also presented for the Society of Automotive Engineers, the Automotive Industry Action Group, MarcusEvans, and the Software Test and Performance Convention.

Additionally, Pries is a principal of Value Transformation, a product development training and cost improvement organization. He is also a founding faculty member of Practical Project Management. E-mail Pries at *kim.pries@valuetransform.com*

Jon M. Quigley has three college degrees, a B.S. in electronic engineering technology from the University of North Carolina at Charlotte and a master's of business administration and an M.S. in project management from City University of Seattle. In addition to these degrees, he holds the following certifications:

- Project Management Institute:
  - Project Management Professional (PMP)
- International Software Testing Qualifications Board (ISTQB):
  - Certified Tester Foundation Level (CTFL)

In addition to these degrees and certifications, Quigley has secured six U.S. patents over the years, with another two in various stages at the U.S. Patent Office, one of which is in the pre-grant stage. These patents range from human–machine interfaces to telemetry systems and drivers' aids:

- U.S. Patent Award 6,253,131, Steering wheel electronic interface
- U.S. Patent Award 6,130,487, Electronic interface and method for connecting the electrical systems of truck and trailer
- U.S. Patent Award 6,828,924, Integrated vehicle communications display (also a European patent)
- U.S. Patent Award 6,718,906, Dual scale vehicle gauge
- U.S. Patent Award 7,512,477 Systems and methods for guiding operators to optimized engine operation
- U.S. Patent Award 7,629,878, Measuring instrument having location controlled display

Quigley won the Volvo-3P Technical Award in 2005, going on to win the Volvo Technology Award in 2006.

Quigley has more than 20 years of product development experience, ranging from embedded hardware and software design through verification, project management, and line management.

- Software engineer
- Embedded engineer (hardware)
- Test engineer

- Project manager
- Electrical and electronic systems manager
- Verification and test manager

Quigley is a principal of Value Transformation, a product development training and cost improvement organization. He is also a founding faculty member of Practical Project Management. He is the co-author of the books *Project Management of Complex and Embedded Systems* and *Scrum Project Management*. Also with Kim H. Pries, he has written more than 30 magazine articles and presented at numerous product development conferences about various aspects of product development and project management. Quigley lives in Lexington, North Carolina. E-mail Quigley at *jon.quigley@valuetransform.com*. Additional information can be found at *www.valuetransform.com*.

Chapter 1

Does Your Testing Look Like This?

1.1 Last-Minute Flailing

Some enterprises regard testing as the final piece of the development effort rather than as a competitive tool. As a consequence, the test team executes critical test plans as the project is closing—just before launch and well after the time when the design and development teams can perform any kind of rational corrective actions. This behavior affords little time for the test engineers to understand the product at a sufficiently detailed level to perform useful testing. Testing groups may adopt a fatalistic approach to their craft by realizing that dilatory sample delivery and schedule crashing is part of their destiny or possibly the result of truly incompetent planning. The situation is aggravated by disengagement between the development and verification groups, as well as a frequent disconnect between project management, development engineering, and test engineering. The seriousness of this situation is illustrated when you hear project managers lament the test group "blowing" the schedule when they refuse the request of testing the system before the constituent system components are even available for test.

We owe it to our customers to provide them with a high-quality, reasonably priced, on-schedule, and safe products. Test engineering is a huge driver for achieving this goal because it is through testing that we reveal the character of our product. If we are professional enough and careful enough, we can cautiously predict the general quality of the product we are about to sell. Intelligent product testing should eliminate surprises in the field.

Intelligent testing means we are not waiting until the eleventh hour to make an acquaintance with our product. The kinds of last-minute flailing we have seen reflect

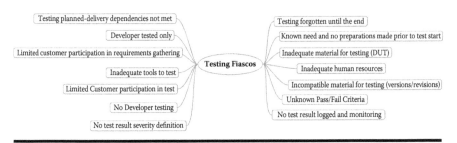

Figure 1.1   Typical testing fiascos.

a kind of cavalier attitude toward the quality of the product that never ceases to arrive as a surprise. In our own positions, we have done whatever we can to implement intelligent and complete testing.

1.2 Fiascos Uncovered Weeks before Launch

Critical dependencies in the product development process often climax during testing—generally at the end of the project. Embedded software often takes a long time to mature; meanwhile, the delivery date rarely changes. We have seen the "we can test late" syndrome in several industries. So, the testing is either much more abbreviated than originally planned or continues after the first products are shipped. Then, the problems are already shipping to the customer. Numerous problems (see Figure 1.1) in the early product delivered to the field can contaminate the customer's perspective of the product indefinitely.

While an organization may think it is saving money by delivering a product on time with minimal testing, the reality is that a poor-quality product may have a long-term impact on the company's ability to receive revenue and improve margins with the product. Additionally, it will be expensive to fix field problems that are not contained in the plant. This will cost engineering time, sales and marketing time, and production time.

1.3 Huge Warranty Problems

Even when an organization tests a product to specification (the minimalist's approach to testing), we generally expect to see subsequent field problems. Some products present difficulties when endeavoring to quantify the field exposure. This ranges from environmental exposure through customer use impacts on the quality. Not only can it be difficult to fully quantify the demands on a product up front, but there can be variation in the product itself and there certainly is variation in how the customer will use the product. These ingredients all add up to a nasty stew that, if

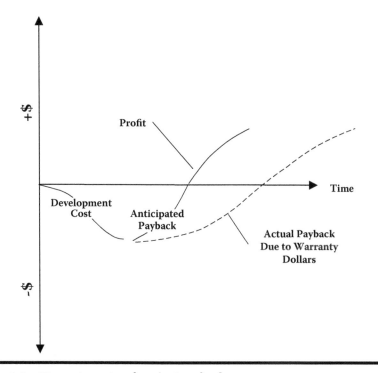

Figure 1.2   Warranty cost and project payback.

taken to the tested-to-specification route, could be detrimental to the future of sales. If the warranty dollars for the product are greater than planned (see Figure 1.2), the profit margins on the product are less than planned. It can be a difficult sell to improve the product margins by increasing the product cost when the product quality is below customer expectations or less than originally touted by the supplying organization.

Testing to specification, especially if each line item of the specifications is performed independent of the other requirements, can present the delusion of successful testing. Suppose an electronic component output was tested to a certain voltage requirement at an ambient temperature of 25°C. Suppose then that the operating environment for the product has a range of −40°C to +85°C. Just because the component is able to deliver the desired current/voltage at the ambient temperature of 25°C does not mean it will be able to do so at the other temperatures (the upper level often being the most risky). The elevated temperatures on electronic equipment require derating after understanding the design work. If we do not follow this level of diligence, it will be largely accidental that the product will be able to deliver the required load current or voltage without some negative consequence upon the product. If this combination is not confirmed, it will be found in the field when we have the customer experience this set of stimuli.

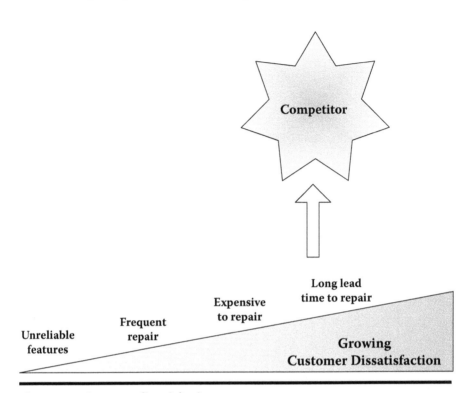

Figure 1.3   Customer dissatisfaction.

1.4 Customer Dissatisfaction

Customer satisfaction is generated, at least in part, by the quality of the product. There are a number of other aspects, but we are discussing testing and how that drives product quality. It is difficult to gauge the impact of customer dissatisfaction in relation to a mediocre product. If our organization is the only supplying organization for the product, then we can still make some progress with the customer relationship for some period, an interval that is diminishing as the customer feels increasingly slighted by the inferior product. If the market size is large enough for the product, eventually some other manufacturers may come to believe they can make a better product. We can then expect our product volumes to decrease as these new, more reliable entries to the market take away our customers with largely the same feature set. Figure 1.3 depicts the process of increasing customer dissatisfaction as well as the shift to another supplier.

Chapter 2

Benefits of Improved Testing

We once heard a product development neophyte say, "Testing is only checking to see if you developed things correctly." While there is some truth to this statement, the reality is that the possibility of getting everything (every task, every interpretation, and every deliverable item that could impact quality) correct in the development of a product is unlikely at best. There are hundreds and thousands of interactions and interpretations required. With transnational enterprises, the work is divided up all over the globe, with conflicting responsibilities, perspectives, and authorities. Additionally, decisions or interpretations are made with limited time and information. It really is not possible to get all these exchanges perfect every time.

Consider testing as a control system (see Figure 2.1). The system is designed to control the results (output) based upon constraints and capabilities of the organization as well as other factors such as risk. This is no different than any other system or activity where we want to control the outcome or at least be able to predict the results.

Making decisions about the economic value of a project is difficult to measure, and adding unpredictable quality issues makes this even more difficult. Whether the calculation is internal rate of return (IRR), net present value (NPV), or return on investment (ROI), the expected economic benefit to the organization is offset by the economic loss of maintaining the product (see Figure 2.2). If this product maintenance carries the additional burden of poor quality, the expected profit may vanish. This loss can be severe if the product has severe defects that harm the customer, initiating legal actions.

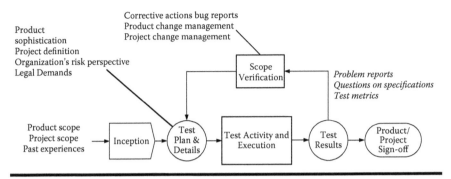

Corrective actions bug reports
Product change management
Project change management

Product
sophistication
Project definition
Organization's risk perspective
Legal Demands

Scope
Verification

Problem reports
Questions on specifications
Test metrics

Product scope
Project scope
Past experiences

Inception → Test Plan & Details → Test Activity and Execution → Test Results → Product/Project Sign-off

Figure 2.1   Testing as a control system.

2.1 Product Problems Revealed Early

Unlike the aforementioned scenario where testing is the last activity in a long line of activities, testing should be integrated into the development effort. The reason is that we want corrective actions that are made by developers as well as feedback from the testing and verification group. This does not absolve the developer from performing an appropriate level of diligence. Figure 2.3 shows the process for the developer's level of verification of the product.

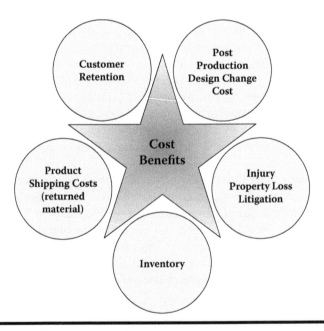

Figure 2.2   Benefits of improved testing.

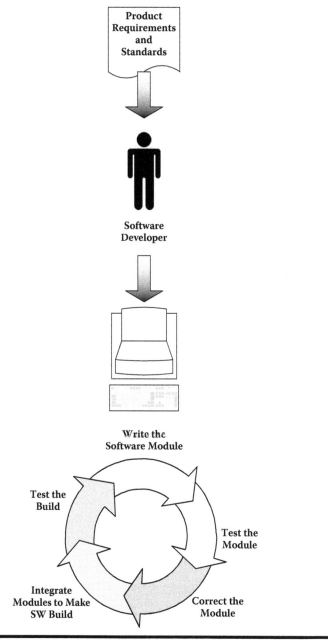

Figure 2.3   Development and verification.

Developer testing typically uses the development tools to verify. In-circuit emulators (ICEs) are used to verify the performance of the software modules. The use of breakpoints and interrupts to monitor the execution of the software within the target system allows for verification to start well in advance of delivery to the test group. Additionally, a developer can use software hooks to test modules. These hooks are interception points in the software that allow the developer to monitor the results. This technique is also employed to alter a variable within the software, forcing conditions upon the system that may be difficult to replicate early in the development process. Another example would be the use of a serial bus to present a view of the key internal information on the product, such as processor load, while operating within the system.

2.2 Improved Reliability = Lower Cost

Improved reliability has the obvious benefit of lowering the maintenance cost of the product—return costs will be less, as well as the handling of these returned parts. Fewer failures mean fewer failed parts to manage, handle, and for which to find the root cause and corrective actions. These activities can take a lot time and resources that do not add value to the supply chain but drag it down.

2.3 Happy Customers

Ultimately, customers do not want a failure. It will not matter whether the product fails due to design, to manufacturing, or to an unanticipated use of the product. Happy customers are ones whose product experiences are not complicated by failures or unpredictable performance, anomalies, and the hassle of returning the product.

2.4 Confidence in a Fine Product

Because the product's actual quality (not estimated return rates) has an impact on the business case for the product, it is incumbent upon the developing organization to understand how much money is at stake due to the product quality. Some of these items are tangible costs—other things, such as *customer perception*, are not so tangible. Nevertheless, knowing we have created a fine product means these quality issues have not eroded planned profit margins to the point where we are losing money on every part made and delivered to the customer.

2.5 Cost-Effective Testing Solutions Not Waiting Until the Last Minute

Cost-effective testing does not mean minimizing the testing or starting the testing at the tail end of the project. Reducing hours only appears to make the testing cost effective. Waiting until the product is nearing completion to start testing is really not very cost effective because we typically incur overtime hours and, more importantly, tired employees may miss a failure mode.

Securing the tools to perform the testing, especially some types of environmental test equipment, can be especially challenging because this equipment can be very costly (e.g., an environmental chamber can run between $20,000 to $30,000). An organization that chooses to purchase such equipment will likely insist on improving utilization to help make the business case. Sometimes, even getting onto the test schedule can be difficult. As difficult as this is when the equipment is in-house, it will be much more difficult when the equipment is out-of-house. In these instances, getting time in the thermal chamber or vibration fixture will be up to luck, hope, and fate. Additionally, it takes time to design any special fixtures that will be needed for the tests. For example, to test a particular product to a vibration profile, you must be able to mount the product on the fixture in some way that represents how the item will be mounted (i.e., orientation of the axis) in the field—all of this with enough rigidity that the fixture is not flexing to absorb energy that should be directed at the product.

These limitations are not just for the physical properties of the product. There is functional testing that will benefit from sufficient time. Regression testing of a product or testing of a complex multicomponent system and all of the interactions will take time just to identify the biggest and most likely risks. This may require special tools and, in some instances, to cover the widest possible functional area, automated testing and simulation activities may be in order. All of this takes time. For the testing to have any positive impact on the product's final quality, there must be time to perform the testing, get the results, and for the development group to take the necessary actions to bring the product performance to expectations.

Chapter 3

Overview

Testing, like good development, is iterative, with close attention to the details. There are multiple loops of testing that feed information back to the development group and allow for corrections and adjustments to the product. Figure 3.1 shows a typical sequence for testing. The physical configuration audit checks out documentation, and the functional configuration audit verifies that the product meets specification requirements.

3.1 Goals of Testing

The objectives of testing, at a minimum, are to make sure the product meets the customer's expectations. This is a minimal approach—to be really successful, the goal of testing is to

- Discover product defects
- Prevent defects
- Contain defects to a single release
- Analyze for statistical release readiness
- Discover product failure limits

3.1.1 Discover Product Defects

One of the last things any organization wants is to discover the problems in the field. A problem in the field requires more work than issues discovered in the customer's factory and the containment portion of the problem becomes unlikely. Once the product is out in the field, containment means that we are inhibiting the

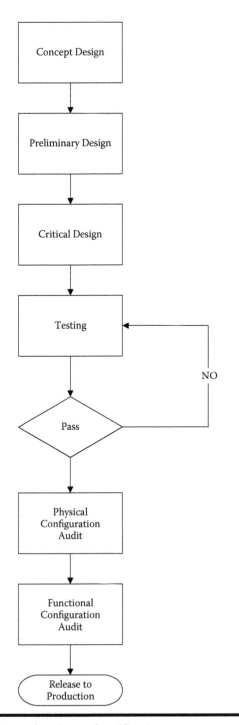

Figure 3.1 Overview of testing and verification.

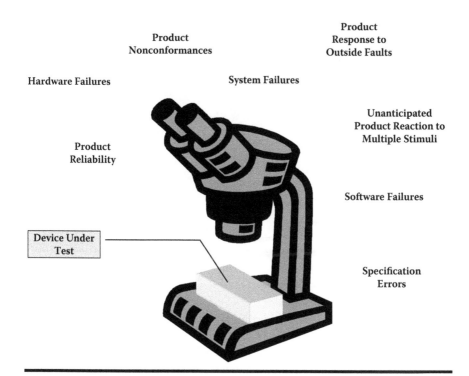

Figure 3.2   Discovery of defects.

exposure—not eliminating it. Often exacerbating the problem is the difficulty in determining the revision or configuration of the product that has the problem. Even when a product is launched with acceptable performance and quality levels, changes deemed minor and left unverified will place the organization in the same situation regarding field impacts.

Additionally, and more importantly, the product defect can have a catastrophic impact on the customer. This can put the customer's safety at risk and result in litigation. Finding field problems is a *lag indicator* of the product quality. By the time you have this information, there can be an abundance of errant product in the field. A *lead indicator* would be the quality of the product under test, the incoming defect rate, the defect closing rate, and defect severity.

3.1.2 Prevent Defects

Effective verification that goes hand in hand with the development process makes it possible to not only discover the defects but also to eliminate them. Development and verification are two halves of the same coin (Figure 3.3). A more systemic view of the effort would use a defect prevention program to attempt elimination of the defects *before* they occur.

Figure 3.3   Development and verification relationship.

Part of that defect prevention could consist of design reviews. The International Software Test Qualifications Board (ISTQB) identifies verification of software, starting with paper verification consisting of reviews of specifications for the product. Critical reviews of the design even in the documentation phase provide the possibility of improving the quality by removing ambiguous requirements of the product even before the testing.

Design reviews that include the test group have a number of benefits. First, the reviews are another way that test personnel acquire an understanding of the product. The test group can develop the most efficient way of testing the product as well as any special tools or fixtures that may be required to execute the test.

Second, if you are testing to requirements, these reviews provide the test group with the opportunity to critique whether or not the requirement is able to be tested or what it will take to test. It is then possible to alter the requirements to ensure appropriate levels of testability.

In addition to design reviews, training programs, the use of models, and process development are parts of comprehensive defect prevention (Figure 3.4).

3.1.3 Contain Defects to a Single Release

We call the process of keeping defects to a single release *defect containment*. We find it unconscionable to permit a known defect to propagate through multiple releases unless it is the result of a business agreement (even then, it is questionable). When the design team makes a correction, we typically test the defective portions of the product first to ascertain if the most recent known defects have departed the software.

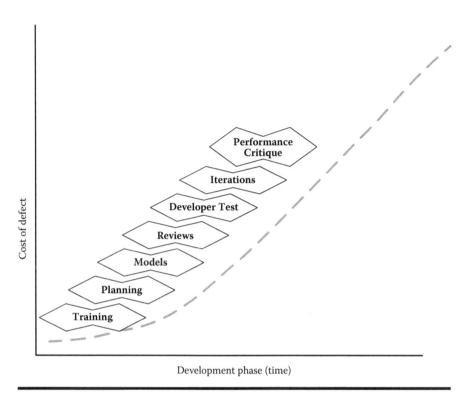

Figure 3.4   Defect prevention.

3.1.4 *Statistical Analysis of Release Readiness*

Monitoring the defect rates over time and over multiple releases makes it possible to spot trends and perform analysis on the test results over time. With software, the standard approach uses a Rayleigh distribution to reflect defect arrival rates. We use the term *defect arrival rate* to reflect the fact that the underlying model is a Poisson distribution; the collection of Poisson distributions creates the Rayleigh distribution (actually a variant of the gamma distribution, but they are related; Figure 3.5). Although the Rayleigh distribution was initially used to model the level of effort in software development projects, it has also been used to reflect the state of defects in the software package. If our data fit the Rayleigh distribution reasonably well, we can make a more confident argument for an appropriate date of release.

3.1.5 *Discover Product Failure Limits*

When performing testing that goes beyond simply verifying conformance to specifications, we provide the customer and the developing organization with a quantitative understanding of the magnitude of the margin between product survival and product failure (Figure 3.6).

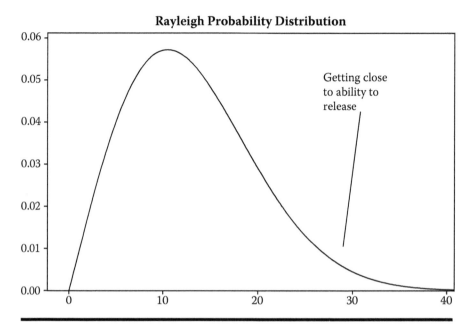

Figure 3.5   Statistical analysis of release.

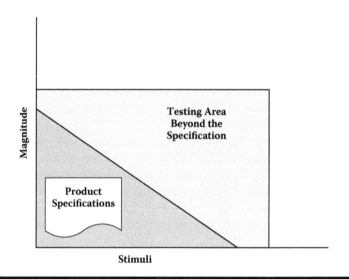

Figure 3.6   Test beyond specification.

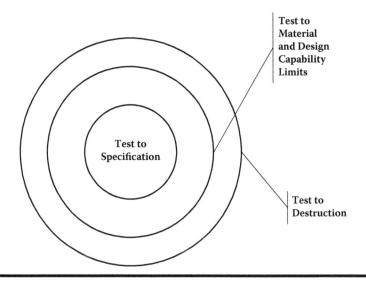

Test to
Material
and Design
Capability
Limits

Test to
Specification

Test to
Destruction

Figure 3.7   Testing to limits.

Testing to specifications merely provides information on the performance of the product within the specified environments. Assessing the significant factors from the potential *external environments* that the product will see in the field can be difficult and time consuming. On the other hand, blindly performing "elephant's foot" tests is sometimes useless because the failure is expected and obvious (an elephant's foot test occurs when we greatly exceed the design limits of the unit under test; for example, having an elephant step on an empty aluminum can).

At every stage, when testing products, we want to:

■ Verify and validate that design limits meet customer requirements.
■ Characterize the failure modes for the product.
■ Characterize the destruction modes for the product.

Figure 3.7 shows an abstract representation of the various limits observed during testing.

3.2 Types of Testing

The method and scope of testing depends on attributes of the product under test. Figure 3.8 shows a typical breakdown of potential tests among hardware, firmware, and software testing. Software development houses will have to test their product on hardware, but they most likely will not directly test the hardware itself. The same might apply to a hardware-only development firm. However, embedded

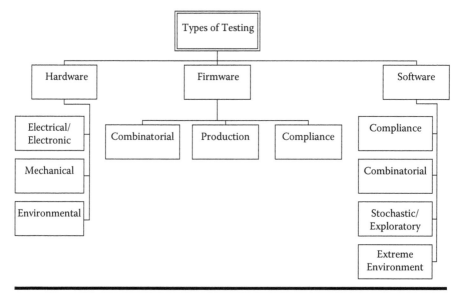

Figure 3.8 Types of testing.

development organizations must know both the hardware and the software as completely as possible.

There are a number of approaches to take to determine product quality and suitability for intended (and possibly unintended) use. To do this, we must ask ourselves a set of questions.

- Which environments will the product *probably* experience?
- Which environments will the product *possibly* experience?
- What can be done in the design to address these environmental impacts?
- How can we best replicate the multiplicity of environments?
- How can we detect faults caused by environmental interactions?
- What level of robustness are we willing to pay for in the product?
- Do we have any safety issues?
- What can we do in the lab or on the bench to get the product to speak to us?

3.2.1 Qualitative

Qualitative tests look for a change in a quality; for example, a color might change. Qualitative tests always involve the use of attributes rather than variables' values (e.g., temperature). Figure 3.9 shows a simple representation of the Kastle-Meyer test for the presence of blood—an archetype for qualitative forensic testing. The test is impressively quick and functions as a decision-support tool by letting us know whether we move on to the much more involved, time-consuming, and expensive

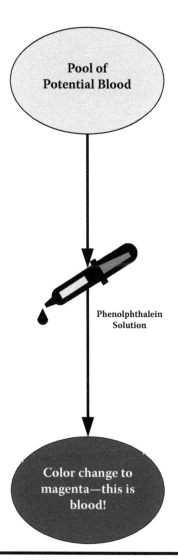

Figure 3.9   Qualitative testing.

testing in a full-featured laboratory. What we see here is exactly what we want from a good qualitative test!

3.2.2 Quantitative

Quantitative tests look for a change in a measurable value—for example, the current in a circuit. Quantitative tests always involve the use of variables rather than attributes (e.g., color). Figure 3.10 shows the sequence of events typical of quantitative testing.

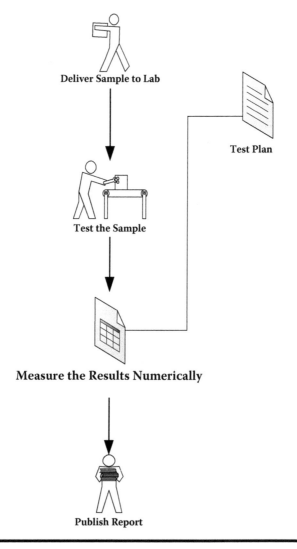

Figure 3.10   Quantitative testing.

3.2.3 Semi-Quantitative (Count Data)

We believe there is a middle ground between truly qualitative testing and quantitative testing—we call this *semi-quantitative testing*. With semi-quantitative testing, we have count data (integers). In many cases, we are counting the number of failures at temperature X. This type of data can also be called *categorical data*, but it is neither quantitative data nor are we really measuring a variable (continuous) value.

We can use nonparametric tests such as Tukey's quick test to analyze the results of semi-quantitative testing with high confidence. The t-test is not as useful—because the values for the t-test should be variables, not counts. Note that many of the familiar statistical tests (e.g., t-tests and z-tests) are intended for use with measured variables; we must use the appropriate categorical statistics when dealing with count data. For example, we might consider the following:

■ Categorical distribution, general model
■ Stratified analysis
■ Chi-square test
■ Relative risk
■ Binomial regression
■ McNemar's test
■ Kappa statistics
■ Generalized linear models
■ Wald test
■ Correspondence analysis

3.2.4 The Four-Phase Approach

At a minimum, we recommend approaching software testing with a multimodal program. The approach is applicable to both software and hardware. The modes are the following (see Figure 3.11):

■ Compliance testing
■ Combinatorial testing
■ Stochastic or exploratory testing
■ Severe environment testing

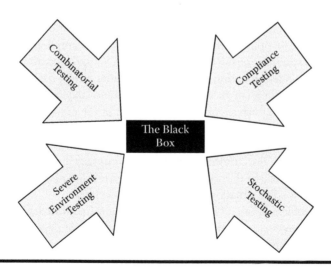

Figure 3.11   The four-phase approach.

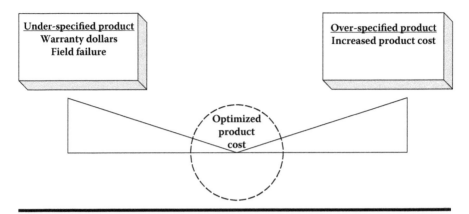

Figure 3.12   Product strength.

3.2.4.1 Compliance

To have compliance testing, there must be requirements to which compliance is demanded. With large commercial customers, these are typically product specifications. These specifications range from the design level all the way to detailed performance demands for the product. Determining the details of these stipulations is critical to getting the right quality of product—particularly the boundary values. Guessing, extrapolating, or intuiting the environment in which a product will be subjected to does not go far in attaining success for the product. Acquiring this information is frequently not a trivial task because we often have derived requirements as well as the more obvious requirements declared in the specification document. The consequences of missing any of this information can cause a variety of negative costs, such as poor launch quality or inadequate field performance. Figure 3.12 shows the boundary values for what occurs to a product when the specification is unbalanced. The effect will also cause the test suite to become extremely complicated.

Products meet specifications all of the time and then fail and cause warranty issues for the customer. Underestimation of the environmental factors has an impact on new products, resulting in premature failures, high warranty dollars, and opportunity costs. These faults can influence customer perception of the product and, if severe enough, possibly lead to product recalls and legal action taken against the supplier, a situation in which everybody loses.

If the product meets the customer's specification but fails in the field, then the customer organization will be the organization bearing the brunt of the consequences of the failure. This is still a situation in which everybody loses.

Even if the problem is not a "hard" failure, functional performance anomalies in ambient environments unpredicted during the development effort can have an adverse effect on the customer.

Overestimation of the demands on the product also adds to cost. An excessively hardened product will have increased material costs. This situation presents an opportunity cost for the organization: excessive investment needed for the product affects the number of projects or products that may be launched.

Additionally, the cost of the product either increases to maintain the same profit margins or the cost does not increase and the margins on the product decrease. Insufficient diligence may reveal that the business case for the product was poor from the start.

3.2.4.2 Extreme Scenarios

What we call *extreme scenarios* are those test cases that push the product to its limits. One electronic example occurs when we provide a microprocessor-controlled device with slow voltage decay. As we pass through the designed voltage limit, the supplied voltage will oscillate slightly—*low-amplitude noise.* This slight oscillation will latch (appear to stop) the microprocessor when the design is not robust. We have also applied random voltages, random currents, voltage slews, varying temperatures (well below freezing to nearly boiling), and other horrors to products as a means of inducing interesting failure modes.

3.2.4.3 Combinatorial

How many combinations of factors can we expect to observe in the field? The answer is that just about any combination can be present in the field, so adequate testing can be difficult to achieve. In some cases, the most critical variable (or "factor") is time. This is especially true when an undesired event is particularly time dependent. How long is long enough? Basically, this question is unanswerable without substantial experimental or field information. If we already have that information, the issue may have already been designed out of the product. Time-dependent failures may be missed because the duration of the testing makes this form of attack more costly than simpler tests. We are always faced with the trade-off between the cost of warranty (or a product recall) and the cost of testing.

Another type of time-dependent issue that can cause substantial heartburn is the intermittent failure and the means by which it is generated. Test standards typically do not provide much guidance for testing that will discover these types of failures. Attempts to accelerate the behavior of the product (e.g., the "fast" odometer) may produce false scenarios that are hardly representative of the reality of the product.

Oftentimes, a form of stochastic testing will elicit such a failure. Stochastic testing occurs when we set up our test stimuli such that they present randomly both in order and in duration. The downside of some stochastic testing is that it may be unrepeatable. These kinds of tests can become repeatable if they are scripted; however, the element of truly aperiodic behavior may diminish.

The choice of stimulus or stimuli can also be based on the desired information we wish to elicit from the product. For example, we can use a mechanical shock test on electronic samples (printed circuit board products) to test the integrity of the solder joints. Thermal shock performs the same function but from the direction of resistance to rapid thermal change.

In this case, the verisimilitude of the experiment is not identical to that in the field, but we can extract information from the sample parts quickly, comparing an older design with a newer design with a minimum of sample pieces. If the newer design is not demonstrably better than the old, we would not pursue further testing with more complicated multi-environmental overstresses; in short, we have saved both money and time.

3.2.4.4 Stochastic or Exploratory Testing

Stochastic or exploratory testing occurs when we allow experienced test engineers to use their intuition, backed by some lunacy, to conduct ad hoc experiments with the product. These tests are not *undisciplined* forays into self-indulgent test execution like some kind of flailing; rather, they allow for growth in the test plan, which, in turn, will lead to growth in reliability of the product. We record all stochastic tests as they are executed. When we discover a test that elicits an interesting response from the product, we add the test to our suite of test cases. In short, we are providing a situation where the test engineer is in a learning environment. Figure 3.13 is a representation of the stochastic (exploratory) approach to testing.

Our experience has been that exploratory testing yields nearly the same amount of discovered defects as those we find using combinatorial testing or compliance testing. Furthermore, this type of testing provides some satisfaction for the test engineer and some welcome relief from the grind of executing the rest of the test cases.

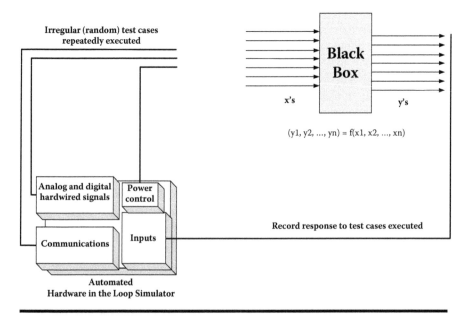

Figure 3.13 Stochastic testing.

3.3 Levels of Testing

There are a number of possible approaches to test and verification. These approaches will depend on the complexity and risks associated with the project and product.

3.3.1 Scale-Up, Scale-Down—The Reductive Approach

Figure 3.14 illustrates the various levels of testing as viewed hierarchically.[1]

3.3.1.1 System

Systems-level testing encompasses the entirety of the product. If the product is to be integrated with a multitude of other modules, then we must account for modular interactions as well as overall system behavior. There is a testing approach, referred to as the *big bang* approach, which essentially gathers all the subsystem parts, puts them together to make the system, and tests them together all at once. Big bang testing is risky if we have never before tested the system or even collections of the subsystems. Figure 3.15 shows some of our test considerations for a commercial vehicle.

As you can imagine, there are some drawbacks to this method of testing. For one thing, it is necessary to sort through all the system interactions to determine any failure and subsequent root cause. For complex systems, this can be very difficult, and a small nonconformance can create a very dramatic impact on the system.

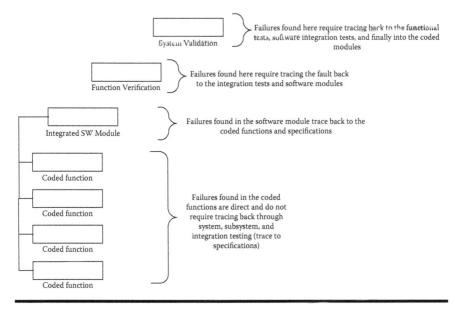

Figure 3.14 Testing hierarchy.

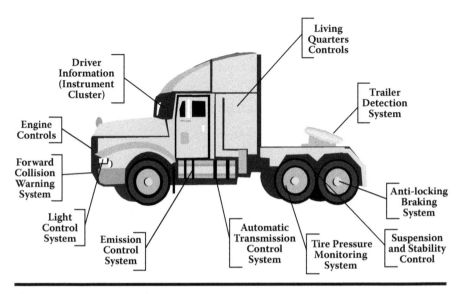

Figure 3.15 System testing.

Additionally, to integrate the various subsystems, or subcomponents of the system, the entirety of the system must be available. This would delay testing until all the constituent parts are available. Experience suggests that this is late in the project and that when integration time comes around, there can be a multitude of problems within each of the constituent parts—this is the first time a look at any part of the system is possible.

Variations in subsystem components and configurations have an impact on the number of test cases. To test to completion every variation possible within a given system can lead to an astronomical number of test cases—so many in some cases that we would not finish the test program in the life of the universe. This situation is sometimes called a *combinatorial explosion.*

The larger the system and the more variation within the system, the larger the number of tests required to verify the product. In large systems that have variation as a function of customer configurations, this situation can require extremely large numbers of test cases, numbering well into the tens of thousands. For these types of tests, automation of the test execution helps reduce this burden. However, there are limits to this approach, particularly when we cannot use a machine and *must* have human involvement in test observation. Prioritization of the potential problems makes it possible to spend appropriate amounts of time on the critical areas.

3.3.1.2 Subsystem

Subsystem testing, as we are referring to here, is an individual device under test (DUT). This can be a collection of hardware and software that fulfills a role within the entire system (see Figure 3.16).

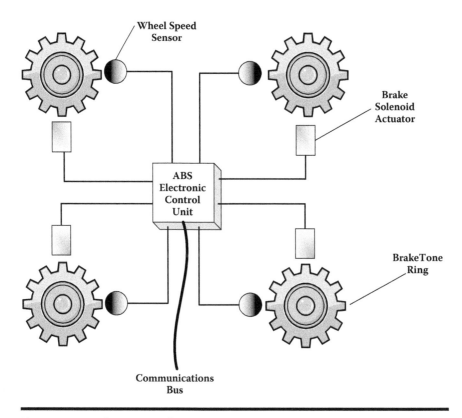

Figure 3.16   Subsystem testing.

We test the subsystem to understand how this portion and version of the subsystem works without the clutter of the entire system interaction. Understanding the limitations of the subsystem will allow us to make decisions about whether and how to integrate this into the entire system. Teasing out the system interaction problems with an unqualified subsystem is more difficult when it is integrated into the entire system.

3.3.1.3 Unit

A unit fits into a subsystem in our hierarchical model. When the unit is a software function and we test that function in isolation, the tests are called *unit testing*. When conducting unit testing, we verify that the unit is functioning properly before injecting it into the larger subsystem of units.

Unit testing is not a lesser form of testing—it is essential to eliminating the discovery of defects during higher-level (and more expensive) subsystem and system testing. We realize, however, that unit testing is unlikely to elicit interaction effects. Regardless, unit testing still makes sense as a relatively inexpensive means of eliminating obvious defects before integration into a subsystem.

3.3.1.4 Sub-Unit

In many cases, we will not use the sub-unit designation. However, we may have a product that exists as a unit (e.g., an instrument cluster) that also has sub-units (e.g., stepper motor gauges). The stepper motor gauges are not components—they are actual sub-units and can be sold separately as service parts.

3.3.1.5 Component

The *component* is the smallest aspect of the system in our hierarchical model. It can be a software module within a device. This level of software testing can be performed by the developing engineer. This person may send expected and unexpected stimuli into the component and monitor the response, making sure the component responds in an expected way and recording unexpected behaviors.

Hardware components can present testing problems, particularly when they are passive electronic components (resistors, capacitors). When an enterprise sells complex, highly engineered electronics, the burden of testing every component can slow the development and manufacturing process to a glacial level of progress. Passive components are usually low-risk components and will not receive much testing, if any. More complex components such as integrated circuits may require verification and validation through test suites that exercise the components.

3.3.2 Levels of Testing

When speaking of levels of testing, we are talking about the hierarchical composition of the product; see Figure 3.17. The approaches to testing can be defined in terms of this hierarchy.

3.3.3 Holistic, Counter-Reductive Approach

3.3.3.1 Component Testing—A Necessary Evil

When we execute tests on components, we will often see only a verification of the data sheet from the supplier. This kind of testing adds little value. If we trust our suppliers and their information about their products, we should be able to eliminate component testing from the decision-making process.

When we have a new supplier or a supplier from a source that has no history with our enterprise, it is prudent to execute some level of component verification to assure ourselves that we are not making a mistake. Other issues will arise:

■ Sample size
■ Specific, significant tests to execute
■ Randomness of parts (parts selected from different lots)

The randomness requirement is difficult to achieve. Many times—especially with electronic products—we receive a set of parts on a tape reel for use in surface-mount

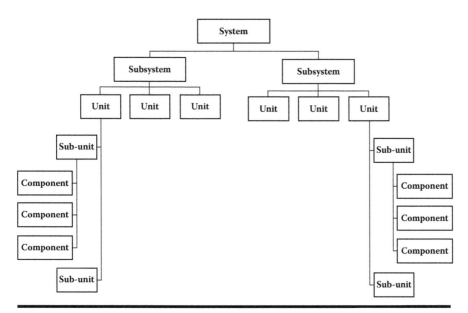

Figure 3.17   Levels of testing.

machines. We can consider the reel as one lot. To use up parts from multiple reels typically interferes with the flow of the surface mount operation so much so that we are looking at considerable expense—not to mention the difficulty of tracking multiple partial reels of parts.

We have also seen cases where choosing the tests that will elicit an issue (discovered later) would be unlikely. In other words, when we have a trustworthy supplier, we will sometimes perform inadequate testing due to our perception of the supplier's veracity. We call this situation *death by data sheet.*

3.3.3.2 Only Interesting Insofar as It Is Part of a System

In all cases where the final product is a *system,* we must consider lower levels of the hierarchy and how they interact with each other. We call this the *holistic viewpoint* in testing. However, a component is never interesting in and of itself but only because it is part of a solution we call a *product.*

3.3.3.3 Multi-Environmental Stress Scenarios

We use multi-environmental stress scenarios for several reasons:

- They provide a more realistic set of stresses.
- We can accelerate product failure.
- We can use far fewer sample pieces and save money.
- We can identify weak portions of the design quickly.

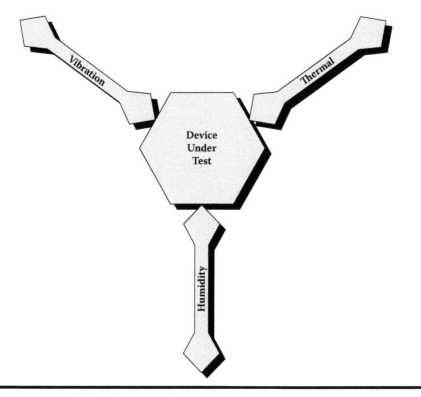

Figure 3.18 Multi-environmental stress.

The interactions of the various stimuli on the product can more closely represent that which the product will experience in the field. Vibration input will exist not in isolation but also with other inputs (see Figure 3.18).

3.3.3.4 Multiple Components, Subsystems

We use multiple components and subsystems so that we can observe interaction effects if any exist. When we test the pieces in isolation, we are unlikely to see such interactions. Figure 3.19 shows a representation of what is occurring as we test the product.

3.3.3.5 Combinatorial Approach to Keep Test Case Count under Control

The combinatorial approach to developing test cases can help us keep the quantity of test cases to a reasonable number while still exhibiting good coverage. With

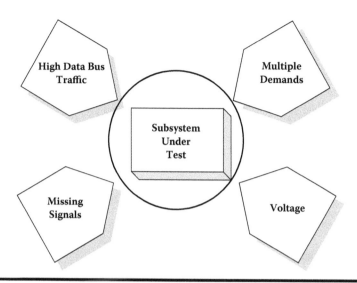

Figure 3.19   Multi-subsystems stress.

electronic products, this approach is excellent for providing a recipe for stimulating digital (zeros and ones) and analog inputs and recording the response.

Note

1. Pries, Kim H. and Quigley, Jon M., *Project Management of Complex and Embedded Systems Ensuring Product Integrity and Quality* (Boca Raton, FL: Taylor and Francis Group, 2009) p. 304,

Chapter 4

Basic Principles

The verification and test group is there to provide some critical and unbiased review of the product. This is used to understand the real quality of the product and make adjustments to improve that quality. When we find a bug or defect, we are in a position to consider whether or not it gets corrected before the product is shipped to the customer. Without this work, the first opportunity to ascertain the product quality would be the customer.

4.1 Looking at the Evidence

Concentrate on What Cannot Lie—The Evidence.

Grissom to Warrick in the *CSI* Pilot

When we test, we are in a position to provide evidence-based results and conclusions to our product development team. In the famous Kalama Sutra, Siddhartha Gautama, the historical Buddha, explains to the Kalamas that they must not accept spiritual declarations as the truth without testing such statements for themselves (see Figure 4.1).

The Method of Science Is the Method of Bold Conjectures and Ingenious and Severe Attempts to Refute Them.

Karl R. Popper

The method of proposing a hypothesis and then testing it is called *abduction* and was first formalized by Charles Sanders Peirce. Karl Popper took the concept further by proposing the principle of *falsifiability*; that is, if we are incapable of

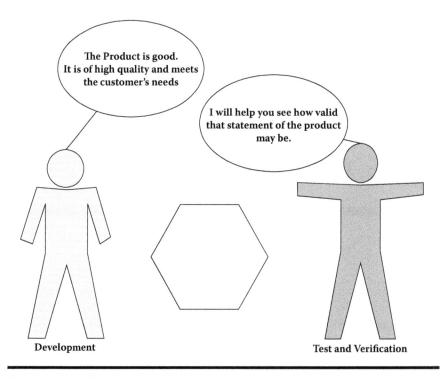

Figure 4.1 Verification principles.

testing a hypothesis, then that hypothesis is effectively meaningless. The principle of falsifiability acts as a practical Occam's Razor to eliminate unverifiable avowals. The approach is not without critics; however, rational use of falsifiability serves to eliminate situations where we are making statements so nebulous as to be meaningless. Figure 4.2 shows the general approach of the scientific method, which is generally a useful approach for any test group.

Usually, a theory cannot be proven true by one test with a positive result, but it can be proven false by one test that disagrees with its predictions.[1]

Note

1. Hartmann, William K. and Miller, Ron, *The History of the Earth* (New York: Workman Publishing, 1991) p. 237.

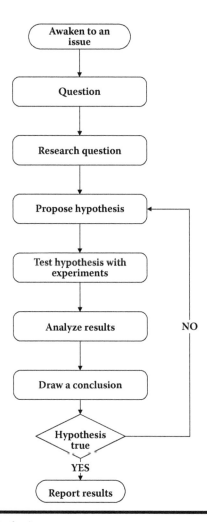

Figure 4.2   Method of science.

Chapter 5

The Question

The question at the heart of testing is not, "How good is it?" It is really, "Where is it bad?" We should be more interested in what does not work rather than futilely trying to prove that our product is defect free. With software, we are generally talking about impossibility, as the number of options possible for testing soon outrace the capability to test those options in the life of the universe. Figure 5.1 is silly but it should serve as a reminder that we consider this question the most fundamental concept in the philosophy of testing!

5.1 Not Phrases

5.1.1 "We Tested and No Trouble Found"

Recently, we were using some software online. The software had what appeared to me to be problems. I contacted the software owner and was informed that I was using the software incorrectly. I read through the description of use and attempted to use it again—only to find the same performance anomaly. This time, I documented the steps, took a snapshot of the output, and sent it to the supplier. The software owner then acknowledged there was, in fact, a bug in the software. I took joy in knowing I helped this software owner clean up his software. On the other hand, I was not on his payroll and as a customer may not like being the test bed—unknowingly—especially if this were purchased software. Figure 5.2 visually indicates the real problem with "No trouble found."

This problem seems to be relatively pervasive—we have personally witnessed this scenario on a number of occasions. There are many instances where a problem is reported, a brief review is performed, or a quick work-through of the function a couple of times finds no evidence of the problem. The engineer will just assume it is a

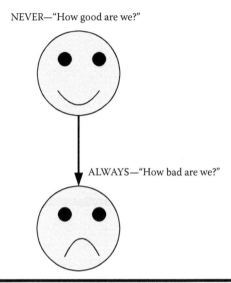

NEVER—"How good are we?"

ALWAYS—"How bad are we?"

Figure 5.1 The question.

reporting issue or inexperience by the reporting party with the hardware or software. Months later the problem is found and subsequently more costly and extensive due the size of the tainted product population.

5.2 Instead

Rather than forgo diligence and tag the customer or reporting party with the "problem," it is better to believe what is reported and take some action to correct the issue. At a very minimum, we have been responsive to the customer (whether internal or external); optimally, we have attempted to avoid—and possibly succeeded—this problem when the field population of the product is much greater. In many cases, getting to the guilty part or behavior can be a serious challenge; however, in the automotive industry, for example, dismantling any part of the vehicle to remove a bad part means that that vehicle is temporarily not out on the road making revenue! The customer is not going to remove a vehicle subsystem for some trivial reason—the product most likely has a real defect.

5.2.1 Assume Customer Observation Correct but May Be Poorly Stated (Erroneous Correlation)

There is something to be said for customer understanding of the product. Especially early on in the product life cycle, the customer may not have full experience with

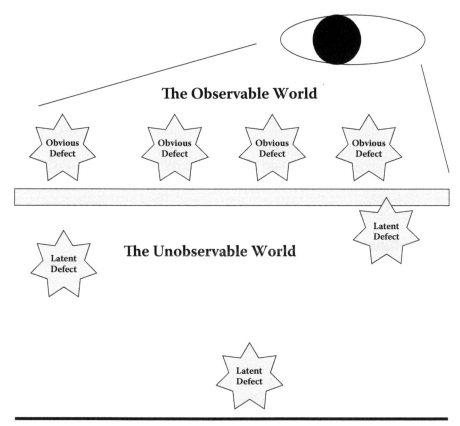

Figure 5.2 Test but don't find anything!

the product and be able to evaluate it appropriately. Every time a customer reports a problem is a customer interpretation—some interpretations are better than others (Nietzsche once indicated that what we really see are interpretations, not facts).[1] People provide knee-jerk responses about the apparent causing event for those things that are more obvious than the inner workings of the product; that is, they point to the apparent failure instead of performing a systematic failure analysis to determine the probable cause of the unsatisfactory behavior. If it is possible to produce erroneous correlation with the obvious, how probable is it to predict cause and effect for something that is opaque to the customer? To do any kind of reasonable root cause analysis, we must enter the exercise with an open mind. One approach to opening our mind is phenomenological reduction, where we suspend our judgment and examine the item of interest on its own merits, shorn of preconceptions and prejudices. This suspension of judgment is called the *epoche*, a Greek word that implies that we are bracketing the item of interest away from conditioned mental formations. Not surprisingly, achieving this level of open-mindedness is difficult. However, we believe it is worth the effort in order to eliminate biases that take us down an incorrect path of analysis.

5.2.2 "What Set of Conditions Could Cause This Event to Occur?"

When we have elicited all we can from the customer about fault information, it is time to proceed further in our analysis. This next step requires investigation of the design to understand how the symptom of failure described could happen by breaking down the hardware and software and the interactions within them to understand the improper behavior of the features to the customer. If the investigator is in the automotive, pharmacy, or food industries, he can resort to an immediate perusal of the design failure mode and effects analysis (DFMEA) and the process failure mode and effects analysis (PFMEA). If our investigator is lucky, he may find pointers to the cause of the issue in these documents.

To be successful, we need to perform a rigorous and systematic critique of the design—with enough follow-up to ensure that any correctable issues have been resolved. Usually, this approach means that we trace the symptom—usually an output—until we discern potential causes. Note that this approach is very close to a logical fallacy called affirming the consequent, where we attempt to find a given antecedent (cause) for a specific consequent (effect). The reason this approach causes problems is that the effect may derive from more than one cause. However, we are suggesting that we compile a list of candidate causes. These possibilities are prioritized for whichever one is the most likely when we think there is enough information to do so. Alternatively, we can use our candidate cause list and induce the observed failures in a controlled environment to test the theory of the root cause (see Figure 5.3). Even with this testing, our conclusions remain vulnerable to error, as a demonstration of *the* failure is not necessarily a demonstration of *the* cause. One method to try and deal with this fallacy is used with electronic parts and has the following steps:

1. Reproduce the observed failure if possible (let's assume success).
2. Hypothesize an electronic or mechanical cause for the failure.
3. Open the unit and test the hypothesis.
4. If the hypothesis appears correct, then repair the part.
5. Attempt to reproduce the observed failure.
6. If we fail to reproduce the failure, we can have some confidence that we did indeed discover the cause of the failure.

Another potential solution is to have failed material sent back for analysis (Figure 5.4). However, we are limited as to what we can do with the failed material unless the failure is a hardware failure. If the product is part of a larger system, then removing the product from the system may remove the stimuli from the *failing*

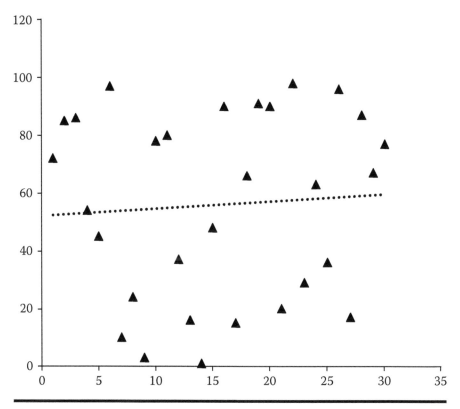

Figure 5.3   Erroneous correlation.

component. If the failure is a hard failure, then review of the failing part as well as review of the nature of the failure provides evidence as to the source of the causing element.

5.2.3 If Issue Not Seen, Assume You Have Not Found the Trigger

It is not time to give up or say, "There is no problem." Customers never want to hear their suppliers tell them it is all in their heads. Time with the customer in the application analysis may be helpful. Finding the scenario where the problem seems to be more commonplace and traveling to investigate the problem where it exists are options in determining the cause. Do not forget that some problems can be related to geography; in other words, we are talking temperature, humidity, rough roads, electromagnetic interference, and other environmental noise. We may even have to resort to a systematic replacement of components to find the guilty part, a task made even more difficult if the "part" is actually software.

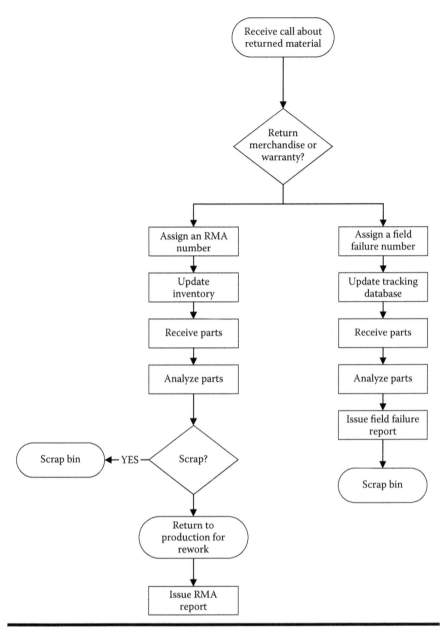

Figure 5.4 Analysis of returned material.

Note

1. Nietzsche, Friedrich, *The Will to Power* (New York: Random House, 1967), p. 267.

Chapter 6

Contradictory Perspectives of Testing

In this chapter we take a look at some contradictory perspectives on testing. These form a type of dialectic that can generally be reconciled with an understanding of the meaning of each perspective and the value that these approaches bring to the testing team. See Figure 6.1.

6.1 Organic/Inorganic

6.1.1 Organic

6.1.1.1 Rhizome Approach

Deleuze and Guattari[1] defined this approach in a seminal chapter entitled "Introduction: Rhizome" in their book *A Thousand Plateaus*. They listed six principles present in rhizome-like structures:

1. Connectivity—the capacity to aggregate by making connections at any point on and within itself
2. Heterogeneity—the capacity to connect anything with anything else, often through the coupling of disparate pieces
3. Multiplicity—consisting of multiple singularities integrated holistically by external relations
4. A signifying rupture—undiminished as a rhizome when being severely ruptured, the ability to allow a system to function and thrive despite local disruption, due to deterritorialising and reterritorialising processes

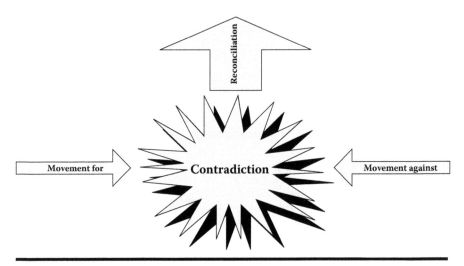

Figure 6.1 Contradictory perspectives.

    5. Cartography—described by the method of mapping for orientation from any point of entry within a whole, rather than by the method of tracing that represents an abstract route, base structure, or genetic axis

    6. Decalcomania—forming through continuous negotiation with its context, constantly adapting through experimentation, enacting asymmetrical active resistance against stiff organization and constraint

One obvious modern example of a rhizome structure is the Internet. Even though the "Internet" is an abstract concept, in practice it functions much like a rhizome, even to the deterritorializing and reterritorializing processes. Real rhizomes grow opportunistically, seeking easy access and sustenance.

6.1.2 Inorganic

MIL-STD-810G, MIL-STD-202G, and the various SAE and ISO test standards are what we call *inorganic testing*. The tests are reasonably repeatable, the recipe is spelled out, and they are used repeatedly by practitioners. For many years, we watched an automotive products laboratory execute these sets multiple times for every project. Yet, the enterprise still released products with issues. In short, the crystalline structure of these standardized tests provided standardized results and little else.

6.1.3 Conclusion

The reality of product testing is that we are most likely to grow our test suite "organically." We don't consider this approach to be evil or unethical but, rather, a simple reflection of the fact we are learning more about the product the more we challenge it.

6.1.3.1 Do We Grow a Test Plan or Do We Impose It?

As we suggested, most test plans arise as inorganic structures based on standards. An automotive test lab might use SAE J1113 for electromagnetic compatibility testing and SAE J1455 for mechanical and environmental testing. These standards present a high level of structure, and the tests are roughly identical from one execution to another. Unless the product is grossly defective, these tests may not detect any problems.

6.1.3.2 Experience Suggests a Combination

A combination of inorganic and organic testing provides an ideal basis for an increasingly challenging test suite. The inorganic portion provides a starting point for the remainder of the test plan. We can still meet customer expectations by meeting their specification requirements and "passing"; however, we can also push the part with the standardized tests and add other tests of our own. In all cases, we want to know under what conditions a product fails and, if we are really aggressive, under what conditions the product will be destroyed.

6.1.3.3 We Grow the Plan from Some Basic Requirements

We can choose an appropriate standard to provide a foundation for our test plan. On this, we can build our superstructure from a variety of sources:

- Historical information from a "lessons learned" database
- A design failure mode and effects analysis (DFMEA) of the product
- What we learn as we go through the testing process with a specific product
- Imagination

6.1.3.4 Plan Grows as We Learn About the Units under Test

The key point in the organic/inorganic debate is to question what we *learn* from a given test or test suite. As we execute our assault on the product, we should be skeptical when we "pass" the tests. A lack of failure or destruction on the part of the product most likely means we just wasted our laboratory time and learned very little about the behavior of the product. Product failures under test will nearly always tell us more about the units under test (UUT) than some abstract status like "passing." A failure is not a disaster if the failure occurs significantly outside the design limits of the product; the same is true of test to destruction.

6.2 Quantitative/Qualitative

After a decade and a half of working in automotive testing, we have rarely seen the lab engineers and technicians consider the use of qualitative testing. We have found that a well-placed qualitative test, however, is a powerful decision support tool.

6.2.1 Qualitative Testing

6.2.1.1 At What Are We Looking?

In general, a qualitative test shows a change in a quality; hence the reason for the name. Sometimes, all we want to know is whether or not we need to progress to a more complex suite of tests. We also want our qualitative tests to be easy to execute and to generate results quickly.

6.2.1.2 Example: Is It Blood?

Since 1903, forensic investigators have used the Kastle-Meyer test to detect the presence of hemoglobin. To conduct the test, they use phenolphthalein because it turns bright magenta in the presence of hemoglobin. The test does not tell us whether the blood is human or how old it is. What the test does do is tell us in a matter of seconds that we are probably looking at blood. The test will exhibit some false positives in the presence of certain vegetables.

Other tests we might do on potential blood are more time consuming and more expensive. Even the portable kit for determining if the blood is human takes more time. If the Kastle-Meyer test does not show the presence of hemoglobin, we are much less likely to progress toward DNA testing. In essence, the Kastle-Meyer test exhibits the behavior we want from a qualitative test: easy to execute, speedy results, and reasonably sensitive.

6.2.2 Semi-Quantitative Testing

We invented this name to cover qualitative tests that produce count data (categorical or ordinal data).

6.2.3 Quantitative Testing

6.2.3.1 Measuring Inputs and Outputs

Black-box testing involves the measurement of inputs and outputs only—we are not using x-ray vision to see inside the sample, whether it is software or hardware. We are not measuring attributes unless we have a means to put a quantitative value on that attribute; for example, we can measure lumens with light and we can measure the emitting bandwidth with color (difficult to do).

6.2.3.2 Inferring Transformations

The transformation of data that occurs in a sample is a mystery with black-box testing and well understood with white-box testing. The choice of one or the other should be a reflection of the testing philosophy and the product requirements. Both approaches can be highly quantitative.

6.2.4 Conclusion

In our experience, engineers tend to prefer highly quantitative results from product testing. What we are recommending is the prudent use of qualitative testing as a method to support decision making and save money. Under no circumstances are we pushing qualitative testing as an end in itself.

6.3 Objective/Subjective

6.3.1 Subjective

6.3.1.1 Customers Can View Testing from a Subjective Viewpoint: "I Don't Want Any Junk"

The inability to provide measurable results to subjective requirements presents the reason why we prefer to see performance specifications that define the required values explicitly. Even when we have customer specification, it does not mean we are not going to add more measurements as well as tolerances to the product.

On the other hand, just because we cannot easily measure a viewpoint or a comment does not mean those items have no value. They may, in fact, be symptomatic of dissatisfaction issues that a customer may have with a supplier. If we realize that we have a potentially negative situation, we may need to develop some measurements that work.

6.3.2 Objective

6.3.2.1 Professional Laboratories, however, Are Encouraged to Approach Testing Objectively

"We measure everything." Measurement becomes meaningful only if we are measuring something meaningful. For example, if we are applying a typical linear staged-gate process to an extremely nonlinear process—due to some true innovations—then measuring the number of deliverable items per gate may not have a whole lot of meaning.

Measuring software productivity by counting new lines of code is meaningless without a paired indicator; to wit, the number of defects encountered in that code. We can sometimes combine metrics like these by creating a ratio; in software development, this ratio is often called *defect density.*

"Here is a picture." Taking a photograph of a failure is not a bad idea; in fact, we can extend the thought by making videos of our tests for subsequent review. If we have positioned our cameras appropriately, we can capture the incidents leading up to the failure as well as any subsequent, less relevant destruction of the failing product.

Of course, the fact that we can let our staff go home while the cameras record potential failures provides a cost benefit to the laboratory. If the customer has an issue with a failure or does not understand the sequence of events with the product, we can package our video on a DVD and provide it to the customer. Of course, we should be using equipment that provides a time stamp for both timing and record-keeping purposes.

6.3.3 Conclusion

Ultimately, our test report will represent a reasonably objective assessment of the units under test. However, we are suggesting that a consideration of the subjective side of testing can help guide us when deciding which tests to include and which to exclude from our test suite. Putting ourselves in the shoes of the customer adds some humanity to our testing approach, preventing us from becoming isolated in some crystalline palace of delusional objectivity.

6.4 Deterministic/Probabilistic

6.4.1 Deterministic Is an Assumption

6.4.1.1 If Bad Is There, Bad Will Be Found

The deterministic testing approach ignores meaningful sample sizes. Automotive testing will often occur with three units under test and perhaps six units under test as we near the end of the development. These sample sizes are too small to detect anything but the grossest of defects. They will detect a probabilistic issue only if the test team is lucky. If the part-to-part variation is high, as is often the case with prototype or early manufacture parts, we won't know if we are looking at representative pieces or assembly ineptitude.

We will occasionally call these kinds of tests *laws of physics* tests because we are detecting situations where we are ostensibly violating the laws of physics.

6.4.1.2 Product Looks Good = Product Is Good

One of the failures implicit in the "we passed" school of testing is the complacency that arises from the illusion of having a good, solid product. This situation provides the impetus for us to repeat that we must test to failure, followed by testing to destruction whenever such testing is feasible. For example, in automotive specifications, the customer will often indicate that they want a part located in the vehicle cab to function correctly up to 85°C. We have seen situations where a marginal component will start failing at around 87°C, which is not a sufficient difference from the required temperature limit.

6.4.1.3 Small Sample Sizes

As indicated, sample sizes of three to six units are inadequate for all but the most obvious defects. If the failure mode we are looking for is an attribute (categorical information) rather than a variable measurement (like temperature), we might expect the required sample size to be in the thousands of units if the event has a low probability of failure (but a high enough severity to pique our interest).

6.4.2 Probabilistic

6.4.2.1 Confidence Intervals

A confidence interval is a statistical estimate of a population parameter (e.g., mean or variance). We calculate an interval that has a high probability of containing the parameter. Confidence intervals are a tool for indicating the reliability of our estimate. Counterintuitively, increasing the confidence *level* actually widens the confidence *interval*—the wider interval is clearly more likely to contain the parameter. The confidence level is expressed as a percentage between 0% and 100%; some common confidence levels are 90% and 95%. When performing our calculations, we should make sure that our model for the distribution of the population is realistic—not all data are normally distributed!

6.4.2.2 Statistical Moments and Distributions

The four statistical moments are as follows:

- *Mean*—the central tendency
- *Variance*—the dispersion of the data
- *Skewness*—whether the data is slopped to one side or the other
- *Kurtosis*—the degree to which the most frequent value is de-emphasized; practically, how much the "peak" has been "pushed" down

The Cauchy distribution has no moments, yet it bears a vague resemblance to a normal distribution. If we are performing reliability testing, we would expect to see distributions resembling the normal distribution primarily in the situation where we are dealing with rapid wear-out of the product. The point we are making is that we need to be cautious about assuming a normal distribution unless that distribution can be demonstrated to be correct.

6.4.2.3 Larger Sample Sizes

In general, the larger the sample, the more likely we are to accurately estimate the population parameters. However, we must recognize that simply choosing a larger sample is not sufficient. It is also necessary to choose randomly and to have some reason for believing that the data are homoscedastic (the variance is the same regardless of where we sample). Sometime, homoscedasticity can be a tall order, as data are often unevenly distributed.

6.4.2.4 Unit Under Test or Product "Life Requirements" Known

Often, the life *requirements* are known but the predicted life is not. The most common approach to this situation is to perform life testing. Unfortunately, many of the models are based on empirically determined coefficients, inverted power laws, and Arrhenius equations. If these approaches are not in accord with what actually happens in the field, we are basically demonstrating and measuring the whole of—usually expensive—nothing.

If our product closely derives from another product, we may be able to apply Bayesian methods to the Weibull distribution, generating a hybrid analysis called *WeiBayes*. Basically, all we need to know is the shape factor with some reasonable confidence and we can generally already begin making statements about the product with the results from three sample parts. Obviously, the more parts, generally the higher the confidence (assuming they are constructed competently in all cases).

6.4.3 Conclusion

We think insufficient consideration is given to probabilistic testing in many test specifications—certainly many of the automotive requirements specify such small sample sizes that only the most obvious flaws will be detected unless we get lucky enough to capture a rarer specimen. We suspect that a more rational approach would be the following:

1. Review field data for a similar product (or products).
2. Examine the expected occurrence listed in the DFMEA.
3. If a fault tree analysis exists, consider probabilities from that document.
4. Try to accumulate any data on reliability we can from the suppliers of the components.
5. Aggregate this information into an assessment of probable failure rates.
6. Determine sample sizes based on probable failure rates and business needs.

6.5 Variables/Attributes

6.5.1 Attributes

6.5.1.1 There or Not There

Attribute testing will verify if a situation exists or does not exist (the binomial condition). We most often use qualitative testing with attribute-based data. Sometimes, we can count the number of times a given condition is present. Count data allow us to use categorical statistics to analyze the situation when we have enough data.

6.5.1.2 Discrete and Small Finite

Discrete data and small finite sample sizes typically make for poor results. By now, this conclusion should be obvious. We only get around this situation in cases where we can use a nonparametric statistical tool like Tukey's quick test, which can analyze a comparison (with high confidence) with as few as six samples of one type (usually the old version) and six samples of another type (usually the new version).

6.5.1.3 Huge Sample Sizes

Sample sizes for attribute data can become enormous; for example, for binomial data we use the following formula:

$$n = \frac{Z^2(\bar{p})(1 - \bar{p})}{(\Delta p)^2}$$

where
Z = Confidence in Z notation
Δp = Desired proportion interval
n = Sample size
$\bar{p}$ = Proportion mean

6.5.2 Variables

6.5.2.1 Measurable

With variables data, we are not *counting* something, we are measuring it. Measurement is one factor that distinguishes variables data from attributes data. In a sense, the measurement provides us with more "information" than we get from simply counting some attribute; hence, our sample sizes are typically much smaller than would be the case with attributes data.

6.5.2.2 Continuous and Infinite

Variables-based measurement occurs on a continuous number line; that is, we have an infinite number of points between any two points A and B. Of course, infinitesimals are not practicably measurable, so the concept has some holes in it. In fact, we say *any* measurement is actually a discrete measurement; however, we are discretely measuring a continuous object.

6.5.2.3 Small Sample Sizes

Sample sizes for variables data can be petite compared with the number required for attributes data (a handful versus hundreds of pieces); for example, we use the following formula:

$$n = \frac{Z^2 \sigma^2}{E^2}$$

where
$\quad Z$ = Confidence in Z notation
$\quad \sigma$ = Standard deviation
$\quad E$ = Error
$\quad n$ = Sample size

This equation makes it clear why we can use rational subgroups of four to ten pieces when computing variables-based control charts (for example, the $\bar{x}=R$ chart).

6.5.3 Conclusion

In most cases, we will want to test using the variables approach, if for no other reason than diminishing the sample size. This statement suggests that we need to be cautious when using terminology like *passed* or *failed*, which implies that we are actually doing binomial attribute testing (pass/fail) when, in fact, we are not doing this kind of testing at all. Attributes testing usually occurs in the following venues:

■ Receiving
■ Incoming inspection (not the same as receiving)
■ Production line product testing

6.6 Continuous/Discrete

6.6.1 Discrete

6.6.1.1 Binary

Binary data (is or is not) are always discrete data. With digital electronics, we simulate binary values using an analog signal. We can usually model binary data with a binomial distribution or a binary logistical regression (on response variables that are binary). We need to ensure that we can readily discern one binary value from another when assessing the results of a test.

6.6.1.2 Stepwise

A stepwise discrete variable may have many values, not just two as in the binary case. The steps of a ladder elevate us above the floor at discrete and stepwise values. If the interval between steps becomes relatively small, we may be able to model the scenario as a continuous function. In any case, the data are discrete and analyses should be performed with experienced judgment.

6.6.2 Continuous

6.6.2.1 Visualize a Number Line with Infinite Points

Continuous data, by definition, do not skip any values; in short, we can use a word such as *infinitesimal.* We cannot actually measure an infinitesimal, but we need to know that the source data is actually continuous. Analog data in an electronic design are *always* continuous—we simply sample values from the data stream. If we are able to sample frequently enough, we can infer the values of the analog source; the Nyquist number tells us that sampling at two times the apparent frequency of the source is sufficient to produce a reasonable approximation of the source.

6.6.2.2 Frequently Intrinsic

Intrinsic continuous values are not dependent on the physical magnitude of the sample. Examples of intrinsic values are the following:

■ Temperature
■ Speed
■ Humidity

6.6.2.3 Temperature Not Dependent on Size

Temperature is an often-used example of continuous data (although we are really sampling). It is also an intrinsic value because it does not depend directly on the magnitude of the substance being measured. We use another value—heat content—when the size of the sample piece is a topic for consideration.

6.6.3 Conclusion

All measurements are discrete because we are always sampling. Continuity is an illusion, much like what we see when we watch a film—which is a sequence of still shots with some kind of *difference* from one shot to another, thereby giving the illusion of motion/continuity. For each continuous statistical distribution, we like to see an equivalent discrete statistical distribution. We can use a normal distribution to approximate binomial and Poisson distributions under special conditions—more because the normal distribution is highly tractable than because of any true relationship with the data.

6.7 Wide-Ranging/Focused

6.7.1 Wide-Ranging

6.7.1.1 Expanded Scope

At some point during a project, we will be faced with the issue of deciding what our scope will be for our testing; for example, are we testing a system or a component in isolation? This consideration is critical because it will determine the extent to which we will set up our testing to reveal interaction effects. Just as with designed experiments, we are nearly always faced with the balance between economy and completeness. In short, we must determine the limits of *sufficient* testing because it is highly unlikely that we will ever achieve *complete* testing, especially with software.

We may also decide that we wish to explore the product with a breadth-first algorithm and follow on with more depth, depending on the results of the first pass. Balancing this combination is a nontrivial matter and requires a high level of technical knowledge backed by experience. As testers, we will always come back to the dilemma:

- How deep is deep enough?
- How wide is wide enough?
- How much can we reasonably afford without adding significant risk?

6.7.1.2 Exploratory

Exploratory testing and stochastic testing are related. With exploratory testing, we allow our test engineers to "forage" for problems. They can use their experience and "gut" to determine their next steps as long as they record everything they do. Any new and valuable tests discovered by exploratory testing should be added to the test case suite for subsequent formal use.

Exploratory testing can be maddening for design engineers because they have no idea from whence the next attack will come—that is the whole idea of using this modality! We have seen engineers beg for us *not* to perform this kind of testing—to no avail. Our philosophy is that it is better to find a defect and delay a project than to meet the schedule and release a defective product.

6.7.1.3 Look for Significant Factors

If we are using design experiments as the basis for our testing, we can conduct screening experiments in the early phases of testing to elicit the significant factors. Screening experiments typically reveal nothing about factor interactions and serve the sole function of identifying important factors for further testing. Once again, we would want enough technical knowledge and experience to back our decisions.

6.7.2 Focused

6.7.2.1 Reduced Scope

Reducing the scope of the testing may allow us to test a specific level of product also known as the stress in great depth. For example, we may allow our tests to run longer to see if we can pick up any wear-out or time-dependent anomalies. If we are performing vibration testing, we might exercise different samples at diverse stress levels to build an S-N curve (also known as the stress/number-of-cycles curve, which is often used in studies of metal fatigue).

6.7.2.2 Usually a Specific Reason for This Scope

If we are performing an in-depth set of test cases, we most likely have technical reason for doing so. Sometimes the reason may be as simple as a root cause search. When we do not ascertain the root cause through analysis of the design, we implement a depth study to see if we can determine the root cause (realizing we are arguing from effects to causes—a dangerous game at best and a logical fallacy).

6.7.2.3 Optimize Significant Factors

If we have run screening experiments (designed experiments for main factors), we can now run more detailed experiments on the main factors we identified. The most potent weapon at our disposal is called *response surface methodology* (RSM). When using RSM, we construct a mathematical entity called a *response surface*. Regardless of whether the surface is planar or nonplanar, we can use an optimizing algorithm (particle swarm optimization, genetic algorithm, steepest ascent) to find the *sweet spot* on the surface.

An alternative approach to RSM would be to use a sequential simplex algorithm. While these are less statistically oriented than RSM, they can sometimes yield an optimization by straight iterations; that is, without using a separate optimization algorithm. The approach uses simple arithmetic and mainly requires bookkeeping and starting with a reasonably good guess at the desired values. The sequential simplex algorithm is not considered a designed experiment in the sense of statistics and orthogonal arrays.

6.7.3 Conclusion

We recommend for most testing that we begin with a wide-ranging approach and perform a deep dive as the results warrant. Other approaches are likely to be less economical and lead to missing critical factors. Basically, we are executing a search algorithm to determine what is significant. For more scholarly discussion of breadth versus depth, read any general artificial intelligence book.

6.8 Many/Few

Often, we will find ourselves in a situation where we have to choose between many sample pieces and a few sample pieces and also between many tests and a few tests. In many cases, the decision is based largely on the cost of doing the test at all.

6.8.1 Law of Large Numbers

We know from statistics that the more information we have, the more likely we are to make correct inferences about the items we are measuring. This scenario is known as the *law of large numbers*. The weak law of large numbers is fairly intuitive; that is, the more information we have, the more likely the expected value is to the theoretically correct value for that mean. The strong law of large numbers indicates that with increasing amounts of data, we are almost surely at the true value of the mean. Do not confuse the weak and strong laws of large numbers with the pseudoscientific law of averages. The law of averages has nothing to do with statistics and everything to do with wishful thinking.

So how do we know when we have enough—that is to say, enough tests or enough sample pieces? Unfortunately, we will often find ourselves in a situation where we do not have enough information to calculate the probability of failure of a given test or a set of sample pieces. We need both the probability and the accuracy of measurement in order to calculate a correct statistical sample size. When we have a test where a violation of the laws of physics is quite obvious, the tendency is to use relatively small sample sizes. Unfortunately, there is really no statistical basis for making this decision. In fact, using small sample sizes will only catch failures that are extremely gross. In automotive testing, we consider this a dirty little secret. Automotive testing is often performed with three sample pieces or six sample pieces. What the small sample sizes mean is that we are unlikely to ever detect any infrequent failures or interactions. Laboratories make these decisions due to limitations in financing and in time.

6.8.2 Correct Identification of Distribution

When we use statistical distributions to model test results, we need to understand that in nearly every case, we are looking at discrete distributions—not continuous distributions. The continuous distribution will typically be an inference from the discrete distribution. In a sense, then, what we are saying supports the use of Bayesian statistics when analyzing our data. In our experience, most testing facilities do not use Bayesian statistics, nor do they use sequential testing approaches similar to those used in the pharmaceutical industry. Sequential testing was first defined by Abraham Wald during World War II, and it was not until after the war that his techniques were declassified. We see no reason to ignore these tools when they are appropriate.

Let's say we have a set of test results and we decide to fit a distribution to our results. We have at least three ways to measure the goodness of fit: (1) the chi-squared approach, (2) the Kolmogorov-Smirnov approach, and (3) the Anderson-Darling approach. Of these three techniques, the Anderson-Darling method is generally the approach of choice. Simply looking at the goodness of fit numbers is an inadequate way to choose a probability distribution. It makes more sense that we understand, as best we can, what the underlying issues really are. In short, we should know something about the mechanism for failure. We have seen reliability engineers arbitrarily choose a Weibull distribution for life data simply because that distribution is one of the most commonly used in reliability analysis. We would recommend that the professional practitioner put more time into understanding the underlying mechanisms than taking the easy way and using a common distribution.

6.8.2.1 Cauchy Loosely Resembles Normal

Sometimes the Cauchy distribution bears a resemblance to a very peaked normal distribution. The Cauchy distribution does not have any moments—it has no mean and it has no variance. Even so, the Cauchy distribution possesses a strong central tendency. We do not want to jump to the conclusion that we have a normal distribution based solely on a strong central tendency; for example, Figure 6.2 (a Minitab plot) shows that a Cauchy distribution looks somewhat like a normal distribution under certain conditions. It is much better if we understand the mechanisms for

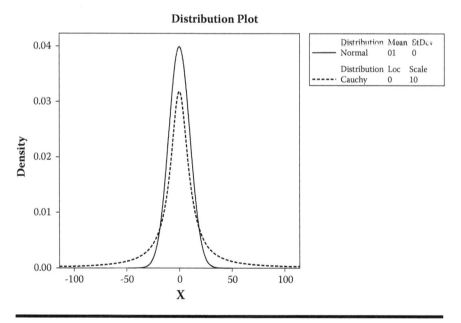

Figure 6.2   Cauchy and normal distributions compared.

failure. We know that a relatively random set of different failure mechanisms will often produce a normal distribution. So, let's go into our analysis knowing that information rather than assuming that information.

6.8.2.2 Which Is Correct?

The best fit is not always the most reliable model for our data. If we are going to do goodness-of-fit analyses to a set of distributions, then we should look at a fairly large set of distributions, say, 10 to 20 distributions. Sometimes, we are going to find that three or four of these distributions have reasonable fits to the real data. Once again, we must understand our mechanisms before we make our selection of a probability distribution model. We also recommend, especially with life data, that we use the empirical Kaplan-Meier fit to the cumulative density function. At a minimum, the Kaplan-Meier fit will help in keeping us honest.

6.8.2.3 How Can You Know?

How do we know we have a good model? The answer is that oftentimes we will not know if we have a good model. However, we can use the Bayesian approach to modify our initial assumptions with subsequent data. While the Bayesian approach gives some statisticians a case of heartburn, we do not see a good justification for ignoring this powerful tool. We think a practitioner is better off making a decision and then applying corrections as new data arrives. Of course, we need to state or define our approach so that the readers of our test reports understand completely what it is we're trying to do. It never hurts to state the assumptions underlying an analysis.

6.8.3 How Small Can We Go?

If we are using count data and our test works out, we can use Tukey's quick test with six pieces of the old and six pieces of the new. Because we are discussing an end count test, all new failures must exceed the test life of all the old failures. Figure 6.3 shows how the end count can be determined.

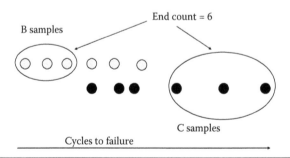

Figure 6.3 Tukey's quick test end count.

We calculate the confidence for a rejection of the null hypothesis in the following way:

- Six end counts gives 90% confidence.
- Seven end counts gives 95% confidence.
- Ten end counts gives 99% confidence.
- Thirteen end counts gives 99.9% confidence.

If we are comparing variables values we have measured, then we can use a t-test with a degree of freedom around four or five with the caveat that the distribution must appear roughly normal.

6.8.4 t-Statistics

We can use t-statistics if we are measuring values and we have small sample sizes. The t-statistic is inappropriate for nominal or ordinal (both categorical) data. When we use baseline-versus-comparison testing, we often use a nonparametric statistic called the Tukey's quick test. The quick test provides good confidence when the two sample groups have a substantially different mean.

6.8.5 White Box

The term *white box* is often used when testing software. This situation is particularly true when the software engineer who wrote the code performs unit tests. When executing unit tests, we nearly always know details about the internals of the software.

6.8.5.1 Hardware/Software Internals Known

With software, we would know the variables, both local and global, and we would know the transformations that occur in the specific function under test. The situation is analogous when testing hardware; with electronics, we would have the schematic and may even have the printed circuit board layout drawings. With this information, we can attack the specifics of the internals of the product to see if we can elicit failures.

6.8.6 Black Box

6.8.6.1 Know Only Inputs and Outputs

We are tying the inputs to the outputs and inferring the effect of any input or combination of inputs on the outputs. With situations where we anticipate timing issues, we might have to consider permutations.

6.8.6.2 Must Infer Transformation

With the black-box approach, we do not have visibility into the internals of the product, whether it is software or hardware. Hence, we must infer the transformation performed by the "box." This approach is not deadly because we are really performing a behavioral characterization of the product much as a behavioral psychologist will analyze for reinforcement during behavioral testing of a pigeon.

6.8.6.3 Should Be Attacked with Combinatorial Testing

As we have indicated, combinatorial testing provides a way to increase the efficiency of our testing by executing tests with combinations of factorial values to determine the effect of the factors on the response. We can use pairwise testing if we do not realistically anticipate any three-way or higher failure modes.

6.8.7 Conclusion

In most cases, we are interested in more data rather than less data. However, we must always balance the cost of performing the tests/experiments against the cost of letting a really foul defect escape our test strategy. As we have indicated, small sample sizes will discover low-probability defects only if we are lucky. We don't like to consider "luck" to be a rational test strategy although we will certainly be thankful any time we see a rare one appear during our testing.

6.9 Structure/Unstructured

The key to this section lies in the lifelessness of heavily structured test standards and test plans. A typical example of such a standard is SAE J1455, the main test standard used in the North American heavy truck industry. The standard per se is fine; however, many heavy truck companies recommend it for design verification and production validation testing. J1455 is not appropriate for design verification with the exception of the grossest design errors. *Production validation* is a poorly defined term in the AIAG Advanced Product Quality Planning document that somehow suggests testing the product one last time will now yield information about its production. Such is not the case. We would like to see both suppliers and customers alike avoid "going through the motions" and, instead, put some thought into their test specifications.

6.9.1 Structure

6.9.1.1 Repeatable

Well-defined structure enhances the probability of repeatability. One example of structure is the use of the checklist. We know in the case of surgery that the WHO checklist reduces the probability of patient death significantly.[2] We provide structure when we issue directives in the following hierarchy:

- Policies
- Procedures
- Work instructions
- Records (a checklist is a record)

6.9.1.2 Known Pattern

We also know that human beings have a tendency to organize concepts into patterns.[3] Furthermore, reproduction occurs naturally through recurring patterns. When we have a pattern that we understand well, we are likely to replicate this pattern. If the base pattern is a relatively optimal pattern, then we are likely to replicate our work with a minimum of errors insofar as we follow the pattern.

6.9.1.3 Consistent

The idea of consistency is consonant with the presence of patterns as well as repeatability. When procedural accuracy is critical, consistency is not the "hobgoblin of small minds."

6.9.2 Unstructured

6.9.2.1 Not Repeatable

It is the lack of repeatability that perhaps scares the earnest test designer the most in these contradictory perspectives. Lack of repeatability may make the use of statistics effectively meaningless in this type of testing. We might say the same thing about the mode-stirring approach to product electromagnetic susceptibility. When using mode-stirring, we attempt to emulate free-space conditions by emanating the signal from antenna plates facing in a multitude of directions. We may not get very high repeatability, but what we do get at a minimum is a very powerful qualitative test. Because the field intensity changes quasi-randomly as the antenna rotates, we may be able to elicit a failure mode that might otherwise remain hidden.

6.9.2.2 Unknown Pattern ... Perhaps More Realistic

Oftentimes in the field, we see an apparent patternless situation or an unknown pattern. The lack of structure in our test plan might be considered a form of randomization—a technique routinely used in designed experiments to provide a natural *block* to systematic external causes. If we choose the randomized approach, we are not so much unstructured as randomly ordered, which is not the same concept.

We might think of the unstructured test mode as similar to a weightlifter working out with free weights, especially dumbbells. By using dumbbells, the weightlifter can increase the degrees of freedom in the exercise, thereby stimulating support musculature. When using a weight machine, the lifter will only exercise his muscles within one degree of freedom. The failure to stimulate support musculature can actually result in muscle damage under extreme conditions, particularly if the individual is a serious or professional athlete. We sometimes see the same situation in testing: We want to increase the degrees of freedom such that we stimulate conditions that would otherwise remain dormant or latent.

6.9.2.3 Inconsistent . . . May Be More realistic

Ralph Waldo Emerson, in *Self Reliance*, wrote, "A foolish consistency is the hobgoblin of little minds, adored by little statesmen and philosophers and divines." A little before this statement, in the same essay, he wrote, "The other terror that scares us from self-trust is our consistency; a reverence for our past act or word, because the eyes of others have no other data for computing our orbit than our past acts, and we are loath to disappoint them." Sometimes this statement is also true with testing.

Perhaps we are beating this concept to death, but it is often inconsistent behavior on the part of the product user/operator that can cause the most egregious failures. The designers may say, "We never planned for that," but such excuse-making is ignoble. Often, we have bizarre and inconsistent behavior left out of the failure mode and effects analyses (FMEAs) required by the following industries just to do business: automotive, pharmaceutical, medical equipment, and food products.

Our goal is not to admit inconsistency through sloppy test case development but to deliberately instigate scenarios that provide inconsistent stimuli to the units under test. One solution to this problem would be to use a driver, either hardware or software, that effectively functions as a strange attractor, never quite retracing the same path through phase space. Another way would be to directly use a random number generator to produce unusual changes in the input (random number generators can be *repeatable* if the same seed is used). We have implemented this kind of approach when supplying voltage to automotive units under test. The SAE J1455 standard suggests that 14.2 V is a nominal value for a motor vehicle operating under "steady-state" conditions (i.e., we are already cruising down the road). In our case, we used a randomizer to move the voltage between the design limits quasi-randomly, thereby providing a more interesting and challenging test scenario for the product.

6.9.3 Conclusion

We do not believe our arguments will drive away structured testing; however, we believe the serious test designer must spend time considering the possibility of using unstructured testing as a means of preparing the product for the unexpected. We remember joking comments made about meteors and meteorite strikes during FMEA exercises; however, in the case of spacecraft, external particle damage is a very real risk and must be considered. Spacecraft also suffer the effects of inconsistent radiation and other natural events. It only makes sense to consider developing a modicum of unstructured tests and test sequences for such a product.

6.10 Ordered/Random

6.10.1 Ordered

6.10.1.1 Repeatable

Tests that are ordered frequently yield nicely repeatable results. However, we should note that designed experiments are generally randomized to form a block against some extraneous factors. We have seen customer specifications that describe a specific order or sequence to the tests they would prefer to see. That means the results from the testing will only apply to that specific order. Randomization of the order would generalize the results—assuming that we have a sufficiently large sample.

6.10.1.2 Reproducible

Ordered testing may also help with reproducibility because our test engineers and technicians become acquainted with the parts and the procedure. The question we need to ask ourselves is: Does the reproducibility we gain by doing the order testing really add value to our results?

6.10.1.3 Intelligible

Because we can usually detect order, we stand a reasonable chance of explaining what we are seeing. However, as always, we need to have an open mind so that we can assure ourselves that any explanation related to order really means something. We have seen at least one experimental case where a piece of test hardware generated such intense radio-frequency emissions that it affected every measuring device in the immediate vicinity. We initially thought we had a journal article; thankfully we realized quickly what was going on and did not make fools of ourselves. We want to always establish the *mechanism* of the order we see.

We can also run into situations where we have emergent order due to a dissipative system. These cases are easy to detect and difficult to explain. Once again, we would expect to resort to an explanation of the mechanism for what we are seeing.

6.10.2 Random

6.10.2.1 Unrepeatable and Irreproducible

True randomness can be difficult to achieve. Practically, we will use a pseudo-random number generator. These algorithms will produce the same sequence of values when we seed them with a specific value and execute them with that same seed value every time. If we are using a seed value like a clock reading or some other reasonably unrepeatable source, the pseudo-random number generator will provide a random set of values and our testing now becomes unrepeatable and potentially irreproducible.

6.10.2.2 May Liberate Unexpected Information from Samples

The beauty of random testing lies in the delivery of unexpected stimuli to the units under test. As noted elsewhere, at some points during our testing, we *want* to assault the product with highly ordered stimuli; at other points during testing, we would prefer to present the sample pieces with a highly varied attack and see how the product reacts.

6.10.2.3 More "Real"

Unexpected stimuli can provide more "realism." We use quotation marks because realism is difficult to define—we find it to be one those "if you have to ask, you don't know what is" kind of items. Basically, we are trying to simulate exogenous shocks—endogenous shocks are more difficult to motivate because they usually imply some element of white-box understanding of the product.

6.10.2.4 Difficult to Implement

The primary difficulty with random testing lies with defining the randomness from the very beginning of the test planning. We can randomize on multiple axes of our attack:

- Magnitude
- Intensity
- Displacement

We would expect significant quantities of nonlinear stimulation as well as nonlinear responses.

6.10.3 Conclusion

In the case of ordered and random testing, we practically have no recourse; we must test both modalities. Extremely ordered onslaughts behave much like fretting fatigue, where a low amplitude but constant stress ultimately persuades the part to fail.

Higher stresses accentuate the attack even more. On the other hand, random stimuli can often simulate reality more closely than the ordered format does. We do not recommend dropping either mode with this particular "contradictory perspective!"

6.11 Nominal/Overstress

6.11.1 Nominal

6.11.1.1 "Meets the Specification"

Sometimes, nominal testing may not even meet the specification. Nominal testing occurs when we do nothing during our testing that challenges the product or service whatsoever. We often see the nominal approach used by neophyte software engineers or lazy engineers when unit testing their product—in essence, they will only conduct the easiest and simplest of tests to verify that the product works without ever trying to find out where the product does not work.

6.11.2 Overstress

6.11.2.1 Exceeds the Specification

Most of the time, overstress testing exceeds the values contained in the specification or standard. However, some standards specify tests that are effectively overstress tests; for example, any one of the six automotive pulse tests (electrical transients) is effectively an overstress test.

6.11.2.2 Characterizes the Part

By driving the part to failure with intelligently chosen tests, we characterize the part under stressful conditions. As always, we want to know when, where, and how the part fails rather than simply passing some abstract test case. Failure is as much a characteristic of the part as success.

6.11.2.3 Set Derating Values

One approach to creating robustness against overstress conditions is to act as if any specific component has design values that are *less* than the actual design limits of the part. This approach is called *derating the part*. By treating components this way, we build in a safety margin and force ourselves to use more robust components from the start of the design.

In order to achieve the same effect with a service, we would need to hire individuals with skill sets or competencies that exceed our needs. When we hire this way, we have people who can flex with unexpected issues, much as the derated part will resist overstress.

6.11.2.4 Know Exactly Where, When, and How It Fails

The skilled practitioner can choose several ways to introduce overstress during testing:

- Step stress, where the stress increases at intervals
- Extended time for testing
- Known harsh scenarios (e.g., near the point of resonance for vibration testing)

We measure where the product fails by looking at failures of multiple parts and ascertaining if a common failure point exists. We know when the product fails through observation or measurement. We know how the product fails by performing a root cause analysis on the failed parts. We should also analyze parts that do not fail to see if we have cumulative damage short of complete failure.

6.11.2.5 Key to Robustness

We have a robust product when it is immune to extraneous influences. We accomplish this goal by decreasing the sensitivity of the product to noise and nuisance factors. One of the most common ways to accomplish this goal is through the use of the designed experiment (DOE). Because designed experiments adjust control factors according to a recipe in the form of an orthogonal array (or some other arrays, as well), we are able to determine which responses are sensitive to variation in the factors and interactions—we use that information to improve the robustness of the product.

6.11.3 Conclusion

By this point in this section, it should be obvious that we tend to favor overstress testing—if for no other reason than we get the results more quickly. We must be careful to avoid generating specious failures unlikely to ever be seen in our universe. Also of concern is to see to it that our teams avoid the complacency that comes with nominal-level testing. We must consider the nominal range for a product to be just another level in a continuum of possible selections for stimulation of failure modes. Nominal stimulation levels are not likely to provide the product characterization we need in order to establish some element of confidence in the robustness of our offering.

6.12 Single Environment/Multiple Environments
6.12.1 Single Environment
6.12.1.1 "Meets the Spec" Revisited

Many times we will receive customer specifications that derive their test requirements from industry standards. This approach is not necessarily deadly, but it is definitely less than optimal. In most cases, the industry standard will specify a relatively simple test with only one major environmental variable. In fact, the automotive standards typically specify boundaries to temperature and humidity to control these sources of noise. The idea is to produce a result that can only be attributed to the factor of interest. Unfortunately, this approach will also eliminate any possibility of detecting factor interactions.

6.12.1.2 Easy to Control

Single environments are generally easier to control than multiple environments, particularly when some of the environmental factors are actually noise (randomly varying and uncontrolled by the test team). If we use high-quality environmental chambers and a good measurement system, we see no reason to continue with the single environment approach other than the comfort provided by the industry standards. In essence, we want to balance repeatability, isolation, and simplicity against realism and significance.

6.12.2 Multiple Environments

6.12.2.1 More Realistic

We touch on *realism* in our section on *verisimilitude*. Physical products and services usually face more than one environmental factor at any given moment. Hence, we find it absurd to expect that a single environment is going to inform us much about field behavior. We find that the greatest difficulty with multiple environment testing lies with choosing which environments and the levels of stimulation necessary to simulate realistic scenarios. We generally get around this problem by testing with baseline versus comparison testing, which allows us to compare the new parts with the old parts that already have some field failure history.

6.12.3 Conclusion

As with overstress, we have a bias toward the use of multiple environments testing simply because the product is nearly always going to see multiple stresses from whatever environment we happen to be using the product in. However, some may question by what means we measure verisimilitude. That is a good question because no metric exists to establish the degree of verisimilitude. Even so, we still think multiple environments are the way to go—at a minimum, it will indicate weak points in the design!

6.13 Compliance/Characterization

Characterization goes beyond compliance testing. Often, our test plans for characterization are supersets of compliance testing; in other words, we contain the compliance testing within the larger test plan. This approach allows us to meet customer expectations with the compliance testing and to exceed customer expectations with the characterization testing.

6.13.1 Compliance Testing

6.13.1.1 To a Standard

By definition, we perform compliance testing against a standard. In the automotive and commercial vehicle businesses in North America, we will most commonly use standards such as SAE J1455 and SAE J1113 for environmental/mechanical and electromagnetic compliance testing, respectively. The rest of the world usually uses International Organization for Standardization (ISO) standards and other documents based on national standards committees. Additionally, we can be required to use military test standards; for example:

- MIL-STD-810G: environmental and mechanical testing
- MIL-STD-202G: environmental and mechanical testing
- MIL-STD-461E: conducted or radiated emissions and conducted or radiated susceptibility

We are not suggesting that test teams throw out their compliance testing! What we are suggesting is that the test teams consider possibilities for liberating defects other than simple standards-based testing. One of us is fond of saying, "The real testing starts when the compliance testing stops." We are making a plea for opening our communal minds and going beyond the go-through-the-motions testing we have seen on myriad projects.

6.13.1.2 Typically OFAT

To date, few standards, if any, exist that specifically describe designed experiments, multiple-environment testing, and other multimodal approaches for attacking the product. That means most of the standards only support one-factor-at-a-time (OFAT) testing, which we have already indicated means that we only discover an interaction effect by fortuitous circumstance. Another issue with OFAT testing is that it is incredibly inefficient, particularly when we desire to test a multitude of actors.

6.13.1.3 Interaction Visibility Minimal

In order to detect interactions among or between factors in an experiment, we need a recipe that is capable of causing enough variation that our analysis of variance can "see" the interaction. This need also means that our experimental arrays cannot be saturated with main factors—they need to allocate columns for the interaction effects. In this regard, the Taguchi approach is somewhat weaker than either classical designed experiments or response surface methodologies. It is possible to detect interactions in nonsaturated Taguchi arrays, but the standard Taguchi approach does not account for the interactions.

6.13.2 Characterization

Because characterization tells us how the product really behaves, every enterprise that executes tests should consider characterization to be a linchpin activity of their validation program.

6.13.2.1 How Do Our Products Perform under Adverse Conditions?

When testing, our goal is never to "find out where the product works" but rather to discover when and where the product or service does not work. With the first attitude, we stop testing when the parts all "pass" the test suite. With the second attitude, we only stop testing when the product disintegrates. If we must progress to extreme testing to yield a product failure, we increase our confidence that the product is robust.

6.13.2.2 Test for Control and Noise Factor Interactions

During product or service characterization, we recommend the use of designed experiments to assess the impact of control and noise factors as well as their interactions. Once we know which factors are significant, we can proceed to optimization of the product if desired. If we are going to optimize, we will most likely use a response surface approach or a sequential simplex algorithm. The response surface approach is elegant and will most likely require some kind of genetic algorithm or some other search heuristic (e.g., particle swarm optimization) to find the sweet spot. The sequential simplex is a search algorithm to begin with; it is possible to use a sequential simplex approach using a simple spreadsheet and rigorous bookkeeping.

6.13.2.3 More Realistic

We consider characterization testing to be more realistic because products will fail— in fact, we drive them to failure. Because all products receive wear, this approach makes a great deal of sense. The only quantitative difficulty occurs when we attempt to estimate the life behavior of the product.

6.13.3 Conclusion

Because good product characterization is a superset of compliance testing, we always recommend the use of product characterization. Most products are not sellable after compliance testing, so the cost differential of testing to destruction is not really greater for characterization. We suspect the cost of *not* characterizing the part is a latent expense that seriously damages an enterprise, both financially and by reputation. Recent issues with Toyota automobiles may reflect a failure to characterize some of the subsystems of the vehicle—at the time of this writing, the jury is still out on the sudden acceleration problem that made headlines.

6.14 High-Level/Detailed

6.14.1 High-Level

6.14.1.1 "Black-Box" Testing

Black-box testing occurs when we only know the inputs and outputs but nothing about the behavior inside the black box. When we conduct this testing, we typically stimulate various inputs and observe the outputs for behavior, expected or unexpected (Figure 6.4).

With black-box testing, we provide a behavioral description of the product. We make no suppositions about the internal functions of the black box. Despite the apparent limitations, we are able to do substantial product testing in the black-box mode. In fact, when a customer uses a product, he will be seeing black-box behavior because he normally has no inkling of the internal workings of the product, particularly if the product contains embedded software.

6.14.1.2 Defects Observed as Improper Outputs Only

With black-box testing, we register a defect as unexpected or improper behavior of an output. In order to conduct our tests, we must have a table that shows the inputs we stimulate and the expected result for each treatment (a treatment is a row in the table). We can enhance the efficiency of our testing through the use of orthogonal arrays, as if we were running a designed experiment—in fact, we *are* running a designed experiment!

6.14.1.3 Can Miss Subtle Issues

We can miss subtle issues during black-box testing because we are not supposed to understand how the product works internally. If we are dealing with complex and embedded software, then we would expect the development team to perform unit tests, wherein we do understand the innards of the design and we test specifically to those design limits.

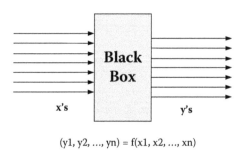

$$(y1, y2, ..., yn) = f(x1, x2, ..., xn)$$

Figure 6.4   Black-box testing.

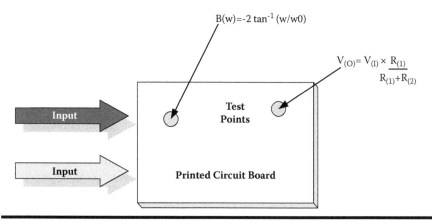

Figure 6.5   White-box testing.

6.14.2 Detailed

6.14.2.1 Time-Consuming

Once we get into the mode where we are examining the inner workings of the product, whether software or hardware or a service, we will increase our testing time and quantity of test cases significantly. The magnitude of the increase will depend on the level of detail we decide on during our analysis of the product.

6.14.2.2 "White-Box" Testing

Instead of looking at a car as the exhaust pipes and the fuel line, we could be looking at the inner workings of the engine. With software, for example, we will have a copy of the code and we will analyze it to see how we can defeat the implementation and any mechanisms put into place to protect the function. With an electric circuit, we will have the schematic and, perhaps, even a copy of the printed circuit board design layout (Figure 6.5). We will know all the values, and we will test by deliberately attempting to defeat the design. Anything less than this kind of aggressive approach will lead to questions about diligence.

6.14.3 Conclusion

High-level versus detailed testing is not really a contradictory proposition but rather an approach to functionally decomposing the problem. In some cases, we may want to attack the product from a black-box perspective to maintain a level of independence and to see the product as the customer sees the product. In other cases, we may benefit from a deep dive into the actual operations of the product and decide to perform white-box testing.

6.15 Growing/Imposing

6.15.1 Growing

6.15.1.1 Progressive Refinement

The idea of progressive refinement allows for evolution of the test suite. We know of a test suite for one product that began with only hundreds of test cases—it now has somewhere in the vicinity of 15,000 test cases, and the test team feels that amount is not enough! (They are correct.) Basically, progressive refinement allows for learning to occur on the part of the test team. As we have said, when we learn more about our product, we begin to add test cases to the test plan. The result of this organic growth is a document that improves as we move through the development process; hence, product versions tested later in the development process receive a more stringent challenge than those tested earlier.

6.15.1.2 Profound Understanding

In addition to increasing the quality and quantity of the testing, a grown test suite also allows members of the test team to become experts on the product or service they are testing. Over the period of the development and during changes afterward, the test staff becomes a significant resource for information about the product.

6.15.2 Imposing

6.15.2.1 Arbitrary Standards-Based Testing

We use the phrase *dropping the crystal* to describe the imposition of standards onto a process. The crystal gives us the illusion of stability and control. Of course, we can always come up with a set of justifications, such as

- We are meeting customer requirements.
- These tests allow us to compare our work with other products tested the same way.
- Our staff knows how to perform these tests.

The justifications are not deadly; however, they also lack vision. We have even seen situations where software developers wanted the test group to *only* execute compliance tests. The software group reacted poorly to results from exploratory (stochastic) testing because the more venturesome form of testing discovered defects. On the test group side, we felt that finding issues with the software in our own development facility was more important than shipping some bad product to the customer and letting the customer discover the issues!

6.15.2.2 Relation to Reality?

Standards-based tests are always abstractions. We could also argue that any testing in a laboratory is also an abstraction. We prefer to think that we can expose the product to differing levels of similitude such that those with greater replication are significantly closer to what we might see in reality.

A typical example of a standards-based test suite is the "load dump." A load dump test simulates the voltage spikes received by a part when a vehicle alternator still functions but the battery cable has been disconnected. This situation may seem odd to automobile owners, but it is not so uncommon with commercial vehicles. The standard defines the voltage spikes and the test group must use a calibrated oscilloscope to measure that the spikes delivered match the standard. The one thing we never do with this kind of testing is deliberately induce a load dump on a vehicle or hardware-in-the-loop setup and record this real event for future playback on the proper equipment.

6.15.2.3 Artificial

We can see, from standardized testing, the artificiality of many tests. This status does not mean the tests have no value, although they may give the illusion that the product is satisfactory when, in fact, the product has some serious issues (e.g., a rare but highly severe failure). We always want to hearken back to the *purpose* of the test: What are we trying to learn about our product? If we see an anomaly, will it be a true representation of product performance?

6.15.3 Conclusion

We expect test suites to grow from a baseline. The baseline will usually be a set of imposed test requirements from a customer (at least in automotive work) or a regulatory agency. That makes the subsequent growth a superstructure to the foundation established by the requirements. We have seen situations where software developers were aghast at changes in the test suite. However, we know that the growth of the test suite truly represents increased knowledge about the product. We can also see improvement over a period of years—the defects that escaped to customers diminished monotonically, reflecting increasing rigor in the testing.

6.16 Bench/Field

We have the option of testing on our bench (laboratory setting) or testing in the field.

6.16.1 Bench

Bench testing (Figure 6.6) clearly has the benefit of generally providing more control over the environment of the unit under test. In some cases, safety requirements may make it obligatory that we perform our tests in a controlled environment.

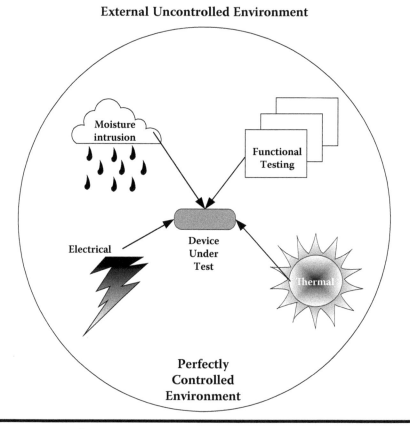

Figure 6.6 Bench testing.

6.16.1.1 Nice, Comfortable Environment

Unless we are absolutely certain that the only way the product will be used is in such a benevolent environment, it is unwise to rely solely on bench testing. One solution would be to use aggressive bench-test environments; for example, we can use high-temperature/high-humidity chambers, near-resonance point vibration, and transient electrical pulses to produce product hell. Figure 6.7 shows factors that influence this undesirable state.

6.16.1.2 Repeatable

A test is repeatable when we can execute it the same way each time we test and get the same results within some specified tolerance. Bench testing is nearly always more repeatable than what we can do in the field, particularly when we are using trained engineers or technicians who have substantial experience with the kinds of tests we are performing.

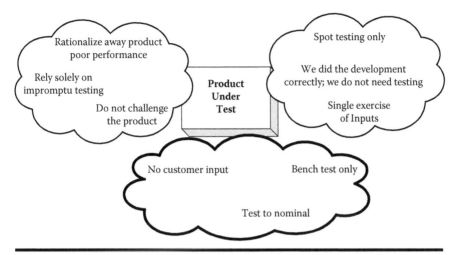

Figure 6.7   Soft testing.

Laboratory certification/accreditation requires demonstration of competence by lab staff. One of the ways of showing this ability is to perform repeatability tests using the same sample, the same person, and the same test equipment and fixtures. We can analyze the results of such a repeatability experiment using statistical software such as Minitab and produce a report that tells us where most of our variation occurs.

6.16.1.3 Frequently Reproducible

Reproducibility is a measure of person-to-person variation when conducting performance experiments with a process or device. In this case, we would use the lab environment to eliminate "noise" variation so that we can more accurately assess the effects of changing engineers or technicians on the results of the test.

When repeatability and reproducibility are paired, we conduct a test called *gage R&R* (gage repeatability and reproducibility). These kinds of tests/experiments are common in manufacturing environments, particularly those that fall under ISO/TS 16949:2002. A full study will tell us where we find the greatest variation and, by so doing, provide us with an opportunity for improvement.

Some sources of variation that we would account for on the bench include the following:

- Measuring instruments (leading to measurement uncertainty)
- People
- Test methods
- Specification or reference value
- Parts

We can use the bench to minimize sources of noise:

- Usage (may see person-to-person)
- Environmental
- Manufacturing variability (may see in the parts)
- Degradation

6.16.1.4 Sometimes Safer

One of us did some research that required the use of potassium cyanide. Potassium becomes a poisonous gas when the pH becomes acidic. In general, we do not perform cyanide experiments in the open air for the obvious reasons. Some regulatory organizations may forbid such activity. In this situation, it makes sense to test on the bench.

In another situation, we had to use perchloric (corrosive and explosive) and sulfuric (highly corrosive) acids to work on metallurgical samples. Both of these substances are highly dangerous. Again, it made sense to work with these substances in a laboratory (the bench). In this particular case, we needed a special chamber in which to do the work that would limit the spread of toxic fumes, explosions, and spills.

6.16.2 Field

Field testing of the product requires installing early samples of the product in the proposed environment for customer interaction. The objective of the testing is to understand how the product reacts in the proposed location. This location can include the most severe applications for the product. This amounts to the worst-case scenario and goes a long way in assessing product capability. In this scenario, a short run (small batch) of production-representative products is delivered and used. These parts must be representative of production in order to isolate design issues from prototype application issues. In the automotive industry, these tests take the form of winter tests and summer tests, usually with extensive road testing. Actual knowledge of the specific stimuli encountered during the test—uncontrollable elements in the test—is not always known or can be monitored. It is not possible to look for what you do not know to look for. Figure 6.8 shows components that go into superior field testing.

Additionally, if the test is performed on "customer" applications—not tested by test engineers—you cannot be sure that faults would be reported with the same critical eye. Features that are not used by the customer or slight anomalous performance may go unobserved. Customer reviews or critiques are not the same thing as verification; however, they do offer opportunities to glean customer feedback on the features offered.

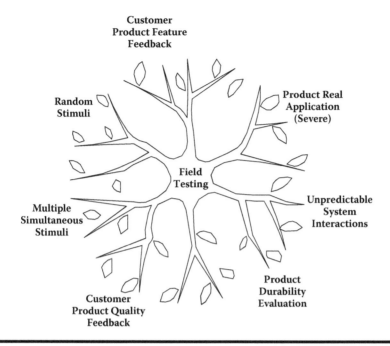

Figure 6.8   Field testing.

6.16.2.1 Often Severe, Random Environment

This randomness is one of the attributes that is often difficult to produce on the bench. Even if we use some kind of random number generator (repeatable, by the way), we may not get the kind of destructive variation we find in the field. Some commercial truck companies will test major changes to a vehicle for a year in an attempt to introduce significant levels of variation. These vehicles will drive through dry and wet, hot and cold, and interactions thereof. Of course, one of the defects of this kind of approach is that we may not see enough variation or that conditions will be so harsh that we will generate outside-the-design-limit failures.

6.16.2.2 Not Easily Repeatable and Not Easily Reproducible

When we move toward absolute similitude by performing, we must make a trade-off between repeatability/reproducibility and the product's performance in its natural environment. Launching products without field testing increases the risk of unanticipated defects arising while the product is in the hands of the customer—a potential path to reputation degradation. In general, we will first test in the laboratory with repeatable and reproducible testing and then follow that with field testing when the product is sufficiently mature that the field testing will have some meaning.

6.16.2.3 How to Provide Sufficient Variety?

If we are going to proceed with field testing, then we need to assure ourselves that we are exposing the product (or service) to as much variety as possible in order to force latent defects out of hiding. Some commercial vehicle companies will test major changes for a year by driving their vehicle in a variety of conditions:

- Hot and wet
- Hot and dry
- Cold and wet
- Cold and dry
- Bumpy roads
- Smooth roads
- Winds
- Many speed changes
- Constant speed

6.16.3 Mixture

To provide suitable test coverage, testing often consists of both bench testing and field testing (see Figure 6.9).

6.16.3.1 What Combination of Two?

The goal when mixing field and laboratory testing should be to respect their relative strengths. We should perform what is unique to each, an approach that reduces wasted time and improves the quality of the testing.

6.16.3.2 How to Control/Monitor/Obtain Good Feedback from Field

Control of field testing starts with the use of rigorous configuration management. By the time the product is introduced to field customers, there should be some confidence in the capability of the product.

Careful selection of those customers who will be part of the field testing is the starting point. You need people who will use the product and be observant of

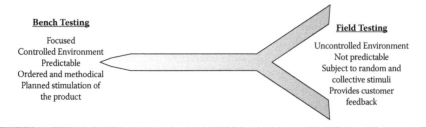

Figure 6.9   Bench and field testing.

product performance under a variety of conditions and not mind that the product being tested may have some defects. Feedback from the customer can be elicited via checklists about key features and expected performance. Another possibility is to provide questionnaires with weighted values to these key areas and performance.

For a sufficiently complicated product, gathering information from the system remotely may be a possibility. One of us has experience with data loggers, devices that record the performance of key variables on the system over time. The product and these devices are periodically reviewed. In some cases, the customer using the test specimen will bring the system back to the developing organization and will be interviewed regarding the product performance. It is also possible to design the recording system with removable media. In this case, the media itself is removed and replaced with blank media. The data logged information on the old media is then shipped back to the developing organization to provide clues on the performance of the product. Finally, with the seemingly ubiquitous cell phone coverage, it is possible to directly offload information about the product directly to the test and development groups.

The choice really depends on the complexity of the product and the risks associated with product failures. Experience suggests that field testing will be unsuccessful without continuous monitoring while expecting customer reports as observable defects occur.

6.16.4 Conclusion

We do bench testing largely because we have to and, to some extent, because some of the standards require controlled environments for their OFAT approach to testing. Field testing can provide a high level of verisimilitude and more realistic stimulation of failures. We have seen some customers reject field testing because they felt it would not be representative—in their case, they were most certainly correct because they were not going to put in the time or mileage to truly flush out field issues. On the other hand, we have seen some customers set a standard of 90 days of field testing for small upgrades and a year of testing for large modifications. They would drive their vehicles all over North America, seeking a variety of test conditions to exercise their vehicle. Another commercial vehicle manufacturer we know of has an incredible test track for stimulating both mechanical and electrical failures on their test trucks. The only real defect to their test track is the lack of severe weather changes (e.g., it does not simulate Death Valley).

6.17 Abstraction/Verisimilitude

In some situations, an abstract representation of reality may be more tractable than a more realistic presentation—one well-known example is that of the subway system in London, England. For many people, the abstract map is much easier to understand than the more realistic version; people want to know the next few stops more than they want to know which river is nearby!

In other situations, we desire a high degree of verisimilitude to keep the scenario as realistic as possible. One of the ways of achieving verisimilitude occurs when we implement multiple stresses simultaneously on the product; for example, by varying the following:

- Temperature
- Humidity
- Vibration
- Mechanical shock
- Transient voltages and currents
- Electrostatic discharge

Even with six stresses, we may not achieve verisimilitude, but we will be closer to the goal than by using one factor at a time. Furthermore, the availability of multiple stressors allows us to implement designed experimentation. Designed experiments are not particularly random—we use a recipe—but allow us to calculate which factors are significant and are a huge step toward producing a robust product.

6.17.1 Abstraction

Abstractions have the advantage of providing a model for testing. The difficulties with abstractions lie in the very meaning of the word: The Latin roots mean to "draw away from." An abstraction may be so far away from the actual product environment that it ceases to have any meaning to the test scenario.

6.17.1.1 Standards-Based

Standards provide a level of abstraction for the test team. While we are not dead set against standards, we feel that the test engineer must remember the limitations of standardized testing. Any sense of "comfort" from passing a standardized test suite is likely to be misplaced.

6.17.1.2 Unreal

The word *abstraction* connotes the unreality of the test scenario. If we are testing a model, we need some level of confidence that the model provides a meaningful representation of the factors under study. In general, we will test a model and then verify the results later on real product material.

6.17.1.3 Pleasantly Repeatable and Reproducible

Of course, one advantage of abstractions is that they are eminently repeatable and reproducible. We can model with

- Matlab/Simulink
- Scilab/Scicos
- Finite element analysis
- Computation fluid dynamics

6.17.2 Verisimilitude

Verisimilitude occurs when we try to replicate the operating environment of the product. The advantages are obvious. The disadvantages are the following:

- It can be expensive to replicate real environments.
- Which environments and how much of each are unknown.
- We never reach complete verisimilitude.

6.17.2.1 Multiple Test Stresses

We indicated one example of multiple stress testing earlier in this section. We can approach the use of multiple stresses in at least two ways:

1. Randomly occurring stresses
2. Designed experiments

With randomly occurring stresses, we are attempting to replicate the aperiodic behavior often seen in nature. We would stimulate the stresses using a random number generator or some other random source (electronic noise) and observe the results. The random number generator approach can make this test repeatable. If we use a true noise source to stimulate our factors, we may find that the test is completely unrepeatable.

With designed experiments, we deliberately stimulate certain factors at different levels based on what our recipe commands us to do. The order of trials can be randomized as a kind of block, but the recipe for each trial is fixed. These tests are relatively repeatable and they will provide us with some useful information:

- The significant factors calculated using analysis of variance
- The magnitude of the experimental error
- The potential for optimization of the design

6.17.2.2 Closer to Reality

We consider multiple stress testing to represent real situations a whole lot more than the more traditional OFAT testing. The nice part about multiple stress testing is that we can use designed experiments and get significant increases in the efficiency of our testing.

6.17.2.3 Sometimes Unrepeatable and Irreproducible

The lack of repeatability and reproducibility is not necessarily deadly to our testing effort, particularly when a specific ad hoc test points us in a new and productive direction. On the other hand, if an ad hoc test sends us on a pointless search because we observed a fluke, then we are wasting our time. Because we have already indicated that a specific test is unrepeatable and irreproducible, we may have trouble detecting whether we are looking at a significant event or an accident. Sometimes, experience with many products can help us.

6.17.3 Conclusion

Our test abstractions provide a repeatable, reproducible playing field. The abstractions simplify our explanations and often make determination of causation much easier. An abstract test can reveal a weakness in the product. The weak spot in the approach, however, is the *relevance* of the observed failure. Increasing verisimilitude also increases our relevance insofar as we have some rationale for claiming that our testing is more real that it might otherwise be.

6.18 Reproducible/Nonreproducible

6.18.1 Reproducible

Reproducibility is a measure of variation from one test operator to another test operator (substitute *engineer* or *technician* as appropriate. Doing repeatability and reproducibility studies allows us to determine the most significant causes of variation in our testing.

6.18.1.1 More Than One Tester/Operator Can Produce Same Results

If we train our people well enough and we can demonstrate reproducibility using the appropriate software (e.g., Minitab), we can show that our results are consistent regardless of the test operator. We would want this information if we were being challenged by a customer or another laboratory.

6.18.1.2 Same Material

Reproducibility studies can use the same material over and over again to eliminate the potential part-to-part variation from the equation. We can also use different parts but only under completely controlled testing, so we can understand the sources of variation.

6.18.1.3 Same Conditions

Ideally, we enhance our chances of achieving reproducibility by verifying that we are testing under the same environmental conditions. If we cannot perform such verification, then we need to be sure we account for noise in our equations before we make any statements about the reproducibility of the testing.

6.18.1.4 Equipment Issues

Some test devices are more amenable to reproducibility than others; for example,

- Antenna testing (radiated immunity) is prone to surprising changes in results with very little movement of the antenna or the sample part.
- Bulk current injection (BCI) uses a clamp on the wire harness of the part and it is easy to show reproducibility with this test scenario.

6.18.2 Nonreproducible

6.18.2.1 Is This Necessarily Invalid?

Nonreproducible results may not be conclusive but they can be suggestive. Often, testing that suggests a new path for analysis is valuable and therefore valid as long we recognize the limitations. In software development, we sometimes call this kind of testing *exploratory testing.* With software, we generally capture what we did, particularly if it successfully elicited a failure mode, and add the results to our standard suite of test cases. Sometimes with hardware, capturing what we did is more difficult, although we should endeavor to record our activities if at all possible.

6.18.3 Conclusion

Laboratory accreditation organizations and standards bodies often want to see proof of at least the capability for repeatability. Certainly, it does not hurt to perform an automotive-style gage R&R analysis to verify competency of the testing staff. Our concern in this section is that we often overlook irreproducible tests as if they were offensive or illicit. As we have noted, at worst they are red herrings but yet, at best, they can be important signals about significant defects in the product. In some cases, an apparently irreproducible defect is actually a low-probability—and significant—anomaly that we want to understand. In short, we got lucky.

6.19 Repeatable/Nonrepeatable

6.19.1 Automated Testing Allows for Repeatability When the Failure Is Truly Repeatable

The use of automated testing involves two major components:

1. A machine capable of executing the tests
2. A script readable by the machine

Usually, the machine is a personal computer or workstation with the appropriate hardware to drive other devices (we use an automotive example):

- Air-driven pistons to actuate switches
- Data buses
- Cameras
    - Optical character recognition (OCR)
    - Gauge pointer positions
    - Gauge overlay colors
    - Telltale light indications and colors
- Servomechanisms
- X/Y positioning devices
- Air pressure controllers
- Digital inputs and outputs
- Analog inputs and outputs
- Noise generators
- Environmental chamber control

Scripts can be written in any language, but some common languages are the following:

- Python
- Ruby
- Tcl
- Perl
- Java
- Visual Basic
- Matlab

6.19.2 Nonrepeatable Testing Is a Serious Problem

Repeatability can have several meanings; for example,

- Metrologists measure the ability of the equipment to measure.
- Testers measure the ability to execute a test again with the same equipment and sample pieces.
- The failure mode itself may have some repeatability.

Aperiodic—sporadic or intermittent—failures are usually a nasty conundrum for the test results analyses. Truly intermittent failures present a probabilistic problem to the test designers. We have seen cases from the field where only one occurrence of a specific was ever observed. A "sample of one" does not mean the failure did not occur but simply that it is particularly rare. We have to ask ourselves follow-up questions regarding the severity and criticality of the failure before proceeding in perhaps fruitless attempts to replicate the failure. Please note that the ability to replicate the failure does *not* mean we know the cause of the failure—we only know about one cause for the failure.

6.19.2.1 In Some Cases, Failure Is Intermittent

When we have an intermittent failure mode, we have a probabilistic condition. In other words, the failure mode will give a strong illusion of randomness, an illusion that may be impenetrable, particularly if the random interval between occurrences is long.

These intermittent failures can also present difficulties because a long-interval failure mode may give the illusion that we have fixed the problem. We might then release a new version of the product and, some time later, begin to see reports from the field describing the same failure mode. These kinds of releases can result in a drop in credibility with the customer because it becomes patently obvious that we never understood the root causes of the failure mode. Given the problem with root causes for a specific failure mode, the fault tree analysis approach to divining failure causes is a better choice in these scenarios. The fault tree addresses some shortcomings of the DFMEA method (single points of failure, no "and" conditions) while adding some shortcomings of its own—it is extremely labor intensive and good, often expensive, software can help the process substantially.

6.19.2.2 In Other Cases, Failure May Never Be Seen Again

One of the worst scenarios occurs when we effectively never see the failure again. We are left in the situation where we cannot replicate the failure in order to do root cause analysis. We may even reach the point where we begin to doubt whether the failure mode ever existed.

We do not feel it is wise to ignore one-off failure modes. It is possible that the situation is telling us that we have a problem with our testing procedure, and it is also possible that the failure mode is real but rare. If the failure mode has a high severity (danger to humans), the issue becomes even more complicated. As we indicated earlier, we must consider severity and criticality for any failure. What is the ethical choice here? If the severity is high, we recommend investigating the test procedures, test setups, and other activities of the test team. We may also want to continue some level of back-burner testing to see if this rare event displays itself again. Regardless, we want to keep records indicating a high level of investigative diligence when dealing with high-severity/high-criticality failures.

6.19.2.3 This Situation Leads to No Resolution

A rare event can ultimately lead to no resolution. However, as we suggested, we can continue with back-burner testing. We should review and update the FMEAs, and we should retain all documentation related to the effort to determine the root cause for the phenomenon. As suggested, we also recommend a fault tree analysis if the severity is extremely high. If the rare event is a high-severity issue, we can expect to be invited to a courtroom at some point, and we need to be able to present evidence of diligence.

6.19.3 Conclusion

Repeatability is important when we want to demonstrate competence at a task—a concept based on the behavior of the executor of the task. However, nonrepeatability can be a product of the nature of the failure, particularly when we do not have a firm grasp on all the factors significant to the behavior of the product.

6.20 Linear/Nonlinear

6.20.1 Linear

6.20.1.1 Easy to Test

Products that behave linearly make testing much easier. We can simply execute our test with linear steps across the behavioral range, record our results, and finish the test with smiles on our faces.

6.20.1.2 Easy to Calibrate

Linear hardware is also easy to calibrate. In fact, linearity is one of the concepts we use in metrology to establish the usable region for a specific instrument.

6.20.2 Nonlinear

6.20.2.1 Difficult to Test

Nonlinear scenarios can present difficulties for testing because we often do not understand the underlying physics of the experiment. However, nonlinearity is not uncommon in the real world. If a product has a specific region of test stimulation where the reactions of the product are very large to small inputs, we would expect nonlinearity, if not chaos.

6.20.2.2 More Difficult to Model

Even when we can test a nonlinear situation, we may not be able to model the experiment successfully. If the occasion warrants explanation via differential equations, we may be dealing with an intractable partial differential equation that may be difficult to solve even with numerical methods.

6.20.2.3 Probably the Most Common in Nature

Nonlinearities are not particularly uncommon in nature. We need only to look at the laminar flow/turbulence transition in fluids (gas or liquid) to view a largely inexplicable physical event.

6.20.2.4 Can Lead to Chaos

Nonlinear physical scenarios may be chaotic; that is, small changes in the setup can lead to significant changes in the resultant behavior. The output behavior may look random rather than regular—under extreme conditions, the output behavior may demonstrate no observable ergodicity. This situation generally occurs when a small input yields a very large output, indicating that the product is sensitive.

6.20.2.5 Dissipative Functions

Dissipative functions were posited by Ilya Prigogine, Nobel Prize winner, and his test group. They observed that, under certain conditions, physical phenomena exhibit some unexpected behaviors. These behaviors occur in *open* systems rather than in the closed systems so beloved by thermodynamics theorists. Also, these systems will generally be way beyond thermodynamic equilibrium. Some common examples include convection, cyclones (especially tornadoes), and hurricanes. More complex examples include Bénard cells (convection cells in fluids heated from below) and the Belousov-Zhabotinsky reaction (a nonlinear chemical oscillator that can produce unexpected changes in color and where particle motion seems to occur spontaneously).

Often, dissipative systems will lead to self-ordering, an emergent behavior. When these issues occur, they throw all the test expectations out the window and lead to confusion on the part of the analytical group. Our point is that we need to be *aware* that these behaviors can occur and have the knowledge to recognize them when we see them.

6.20.3 Conclusion

We suspect the world of being more nonlinear than linear. We are not suggesting a wholesale leap into nonlinearity. What we are suggesting is that we remain *aware* of these nonlinearities and avoid forcing linear solutions onto nonlinear problems. Whenever our observations do not seem to "fit," we may be seeing a nonlinear effect.

6.21 Fine/Coarse

6.21.1 Coarse

6.21.1.1 Very High-Level Testing

Coarse testing may be very high-level testing. For example, we only need to see one failure to know that the product most likely has more failures lurking in the design. Regression testing is often a kind of high-level, coarse testing created to show that no major issues were introduced with a software code change.

6.21.1.2 May Be Used as a Rough Prelude to Making Decisions Regarding Testing at a Finer Granularity

A coarse test can often function much like a qualitative test. We do just enough work to be able to make a decision as to whether we will proceed to a finer test—which is usually also more expensive.

6.21.1.3 Early Development Confirmation

Sometimes we use coarse testing as a means of doing a "quick-and-dirty" test so we can make a decision immediately. We saw an example recently where a test engineer was looking at a problem with a gauge pointer (needle). The black pointers could be hit against a table without the counterbalance falling out; but when he tried this experiment with a red gauge pointer, the counterbalance fell out with almost no effort—suggesting a significant difference between the sample pieces. This example is an extremely coarse test, but it was of sufficient value for a project manager to contact the supplier for a corrective action to their product.

6.21.2 Fine

6.21.2.1 May Miss the "Forest"

When doing "fine" testing, we will use highly accurate instruments and we will be considering the position in the test hierarchy of the units under test. We find that one disadvantage of the fine approach is the failure to stand back from the product and see the bigger picture of what occurs. Furthermore, some test events do not merit a fine approach; for example, catastrophic destruction of the product is obvious—in fact, the test might harm the measuring devices.

6.21.2.2 Need to Understand and Balance the Two Levels

We suggest moving back and forth from fine to coarse and back to improve our understanding of the test events. We do not want to consider only a tiny event when, in fact, we have a cascade of failures occurring.

6.21.2.3 Consider a Continuum

The fine-to-coarse testing is a continuum, with change across the continuum. We suggest that these approaches be planned so we can take the opportunity to use the appropriate measuring instruments for the test—avoiding any damage to expensive equipment.

6.21.3 Conclusion

The idea of fine and coarse gradations is related somewhat to the idea of breadth and depth as well as qualitative versus quantitative. Sometimes a truly coarse test case or set of test cases will better illustrate a defect than a much finer (and usually more expensive) approach. We can also be subject to a forest/trees conflict. Hence, we suggest a mixture of these approaches, generally starting with coarse and moving toward fine as we proceed through the test suite.

6.22 Combinatorial/Exhaustive/Stochastic

In this section we discuss strategies for handling the plethora of possibilities when designing test cases and test suites. When software testing, we expect to see a combinatorial explosion as the number of potential test cases exponentiates due to the inherent complexity of the software. Hardware testing can also present an astronomical number of possibilities.

6.22.1 Exhaustive

We sometimes call the exhaustive method the *Soviet* method because exhaustive testing is often poorly conceived; that is, we are not using our brains to pursue test efficiency.

6.22.1.1 Impossible to Achieve

In the case of software testing, exhaustive testing is effectively impossible to achieve. We assume the code is realistic production code that has external inputs and outputs, selections ("if" statements), and repetition (loops). Nontrivial code will yield uncountably finite possibilities in very short order.

6.22.1.2 Inefficient

Exhaustive testing can be incredibly inefficient, particularly when we consider that we can usually generate high levels of coverage using orthogonal arrays and other designed experimental approaches. The design experimental approach allows us to vary the input factors systematically such that we can observe the behavior of the output effects. We can choose among full-factorial approaches that look at all possibilities or choose to use super efficient Greco-Roman hypercube approaches.

Additionally, if we have existing data or assumptions based on customer knowledge of the product, we can use statistical approaches to put our test case coverage where it is most likely to be seen. In short, we may have some features that are highly unlikely to be activated by the customer. We understand that this approach means that the test team is willing to assume some risk to improve the efficiency of the testing. That risk is probably going to be there anyway because all but the most trivial situation cannot successfully receive exhaustive testing in the lifetime of the universe.

6.22.2 Stochastic/Exploratory (Random)

Stochastic testing is sometimes called *exploratory testing*. Using this technique, our testers are allowed to use their personal experience with software and hardware products to stimulate regions previously untested with formal test cases. If we elicit a really nasty result, we can add this new weapon to our test case armamentarium.

6.22.2.1 Adds a Touch of Reality

Stochastic/exploratory testing can add a touch of reality by simulating the behavior of real users. For example, we can toggle a hardware switch with a light touch or a heavy touch unpredictably. The results of a hesitant touch can be interesting indeed, particularly if this situation is supposed to be managed by both software and hardware.

6.22.2.2 May Not Find Issues

In some situations, exploratory testing may find no issues. This result could mean:

- We did not explore the sample space sufficiently.
- The product is, in fact, robust.
- We are providing insufficient variation during the exploration.
- We may need a machine method that "learns" how to stimulate the unit under test.

6.22.3 Combinatorial

When we recognize that the interactions of stimuli can have an impact on our product, we will perform combinatorial testing. The goal with this is to define multiple stimuli that the product will probably experience.

Run Order	Factor 1	Factor 2	Factor 3	Factor 4
1	−1	1	−1	−1
2	−1	1	−1	1
3	1	−1	1	1
4	1	1	1	1
5	−1	−1	1	−1
6	−1	1	1	1
7	−1	−1	−1	−1
8	−1	−1	−1	1
9	1	−1	−1	1
10	−1	1	1	−1
11	1	1	−1	1
12	1	1	1	−1
13	1	−1	1	−1
14	−1	−1	1	1
15	1	1	−1	−1
16	1	−1	−1	−1

Figure 6.10   Design of experiments.

6.22.3.1 Example: Design of Experiments

Designed experiments provide opportunities for increased efficiency and coverage (see Figure 6.10). A typical designed experiment will use orthogonal arrays (balanced) to exercise different levels of multiple factors with each run or treatment. In essence, the user will often generate such an array using appropriate software to provide the "recipe." The array can be set to test at various levels, depending on the needs of the test group and the end customer. The designed experiment provides a rational, mathematically sound technique for covering inputs and measuring responses. If used during the design phases, designed experiments can produce a transfer function using multiple regression that will allow for optimization using an approach like a genetic algorithm.

Efficiency is one thing, but the other truly wonderful thing that designed experiments can capture is the presence of factor interactions. Interactions cannot be discerned using the one-factor-at-a-time approach where we hold all but one factor constant and vary that factor through some range. In fact, the one-factor-at-a-time approach is one of the most inefficient means to practice testing.

Another benefit of the designed experiment approach lies in the ability to use tools that can optimize desired behavior of the product. Sometimes, we desire a signal to be sensitive because that sensitivity can provide a high measure of control; in other cases, we may desire some function or feature to be insensitive, making it robust to external influences. Optimizing approaches can include particle swarm optimization, genetic algorithms, and other fitness function search algorithms. All these approaches attempt to find the system optimum rather than a local optimum, thereby providing the very best choice.

6.22.3.2 Example: Pairwise Testing

We use pairwise testing as an efficient method for providing coverage of test situations. The advantage is that pairwise testing can be extremely efficient, easily reducing the number of test cases by a magnitude. The drawback is that we will only detect issues that involve paired interactions. This drawback may not be deadly to our test plan because pairwise failures are relatively common. Basically, we are faced with the same situation when conducting designed experiments, where we often ignore interactions of order greater than two because three-way and higher interactions are relatively uncommon. If we are wise about our test regimes, we can also use a three-way test generator in much the same way we use a pairwise test generator (Allpairs from James Bach).

6.22.3.3 Very Efficient

If we take five factors at three levels for each factor, then we have 243 possible combinations, whereas if we use pairwise testing, we can reduce the test trials down to 11 test cases! Of course, we have a trade-off—with pairwise testing we will not see three-way or higher interactions. The question we must ask ourselves is whether we anticipate the existence of three-way or higher interactions.

6.22.3.4 May Miss Some Interactions

Any designed experiment or combinatorial testing array that is saturated with main effects will not be able to present interactions meaningfully. Experimenters call this *confounding* or *aliasing*.

6.22.4 Foreground/Background: Foreground

6.22.4.1 Gestalt Concept

Whenever we perceive, we direct our intention toward an item in the foreground. We see the foreground in greater detail than we see the background. The foreground has greater clarity and is more vivid. Hence, we tend to be foreground biased, both in vision and in our thinking.

6.22.4.2 What Catches Our Attention (and Our Intention) Is Not Necessarily All There Is

We need to remain cognizant of the totality of not only our testing environment but also the field environment. To focus on the foreground while forgetting the background can be dangerous at worst and mistaken at best. We can either redirect our focus or attempt to enlarge our focus. The phenomenologists use a term *horizon* that covers this situation a little differently than *background*. A phenomenological horizon has no limit. We do not completely consume our perception/experience of things regardless of the number of times we reflect on them. Some new horizon or modification of the existing horizon occurs whenever we think the current horizon has receded from our consideration. The experiential horizon is why we say "testing is infinite" with some level of gravity.

6.22.5 Foreground/Background: Background

6.22.5.1 Gestalt Concept

Background correlates with foreground in the Gestalt approach. Gestalt methods deal with perceptions, as opposed to learning theories such as behaviorism—although they both fall under the general heading of epistemology, or how we learn. One of the key Gestalt concepts is the so-called principle of totality wherein our perceptions must be reflected on globally and simultaneously so that we can comprehend what we perceive as part of a system of dynamic relationships. Buddhists incorporate some of these concepts in the idea of the five aggregates: *form, feeling, perception, mental formations,* and *object consciousness.* The five aggregates are key to understanding delusion and so they can function in our concepts of testing.

6.22.5.2 Frame for Our Attention/Intention

As each horizon (or background) arises, it appears to our consciousness as the ground for the observed phenomenon (e g., failure mode). When we are able to horizonalize, each perception retains an equal value while we endeavor to reveal its nature. Prejudgment is the path of missing important information, even though these prejudgments often seem as if they were shortcuts to a quick conclusion to the analysis.

6.22.5.3 May Contain Critical Information

If we horizonalize our consideration, we open our minds by treating all information from our testing (and warranty data) as if it had equal value, at least initially. In other words, we make a deliberate effort to treat the situation on its own terms rather than ignoring portions of the experience due to our preconceptions. This approach is common in psychology, but we think it also has a place in our reflections on our test results.

6.22.5.3.1 What We Think We Know Impedes What We Need to Know!

The essence of bracketing is to open our minds so that we can truly see what is before us (and sometimes what is behind us).

6.22.6 Conclusion

Exhaustive testing is an oxymoron in a sense, because it is highly unlikely we will approach exhaustive testing at any reasonable cost level. Furthermore, how do we *prove* our testing is exhaustive? We use combinatorial approaches to achieve reasonable coverage more efficiently. We use stochastic (exploratory) testing to better use the experiential skills of the test engineer.

6.23 Focused/Fuzzy

6.23.1 Focused

6.23.1.1 Easy to Perceive

We have already discussed some perception issues in the previous section on foreground and background. A similar concept is that of focus; for example, we are focused on the foreground. The focus is visually sharp, which again can lead to perceptual bias that leads us astray from the real problem.

6.23.1.2 May Not Be Analytically Possible

In some cases, we may have difficulties focusing. The preceding statement can have several meanings:

- We are distracted by administrative trivia of the daily job.
- We may not have enough information to assess a situation clearly.
- Our instruments may not be accurate or precise.

6.23.2 Fuzzy

6.23.2.1 May Represent the Situation Better

The advent of so-called fuzzy control indicated a realization that some problems cannot be readily modeled. The idea of fuzziness allows us to treat a problem as if it were composed of contiguous regions and also of mixed regions (hence fuzziness). We can also analyze test results in this way if we must. We would look at the problem to see if we are looking at simultaneous root causes, some of which may be mixed.

6.23.2.2 Piecewise Understanding/Implementation

We have no rule book that says we must perceive a situation holistically, although we generally prefer to do so in order to provide context for our observations. We may be faced with a scenario where we must construct our schema from bits and pieces, much like the motley garment of a harlequin. This approach to patchwork schemata is analogous to the patchwork "lozenges" used in fuzzy control.

6.23.2.3 May Be Purely Empirical

Our fuzzy analysis may not fit an analytical theory very well, but it may provide enough information to allow us to make decisions about the product.

6.23.3 Conclusion

We usually resort to fuzzy approaches when it is unlikely we will ever achieve an analytical understanding of our universe of discourse. We find nothing intrinsically wrong with focused approaches if they are the end result of a reasonable search algorithm rather than the "I know, I know" situation we discuss many times in this book.

6.24 Instant/Enduring

6.24.1 Instant

6.24.1.1 What Is at This Moment

Instantaneous phenomena are transient phenomena. We need to have measuring equipment that can detect high-speed changes in the signal. This requirement implies a sampling rate at least twice that of the frequency of what we are looking at (Nyquist rate). Note, however, that this rule of thumb becomes less meaningful if we do not have a periodic (regular) signal.

6.24.1.2 May Be the Situation Now but Very Transient

In some cases, we test by deliberately presenting short-duration transient signals of high amplitude to the units under test. These tests measure the ability of the protective hardware to block damage from these signals. These signals can be electronic or mechanical.

6.24.1.3 Reality Has an Infinity of These

Reality has an infinity of instantaneous moments. We are limited by our ability to sample and it can be the case that our measuring equipment will miss an extremely short-duration transient.

6.24.2 Enduring

6.24.2.1 May Be an Illusion of Continuity

We know that illusions of continuity occur because we see them every time we watch a film, which is really a high-speed merge of discrete images. With this knowledge, we need to analyze our warranty parts and test parts as both a continuity and as a collection of discrete events.

6.24.2.2 Could Be Comprised of Similar Instants

Our movie metaphor suggests that our perception of things is really a process of discrete instants, a thought that led David Hume to question some of our ideas of causation. This concept applies to test parts and field parts that we often treat as if they were utterly a continuity when, in fact, they are composed of discrete quanta of activity. Hence, we find it important to consider the time constant when analyzing parts. We have seen some Monte Carlo models that treat physical events as if they occurred instantaneously.

6.24.3 Conclusion

We must consider the time factor every time we test a complex and embedded system. We have seen too many Monte Carlo models of so-called worst-case scenarios that ignore the time factor, thereby producing distorted results leading to incorrect conclusions about the final behavior of the product. We also expect our test engineer to consider the effect of intense but short-lived transients, as some of our electronics cannot handle these kinds of intrusions. On the other side lie the long-term effects of stimulation. We are often constrained from seeing the failures we need to see by monetary considerations, the unreality of speed-up code, and lack of time. Unfortunately, long-term failure modes will often find their way to the field, where we are unable to effect improvements until we have a large enough collection of failed parts to be able to say anything about a root cause.

6.25 Inside System/Outside System

6.25.1 Inside System

6.25.1.1 Requires Defining the System

Figure 6.11 shows the crux of the system problem for an enterprise—which level really is the system? How much of the system must we consider to help define the scope for the issue under consideration? If we scope our system too large for the problem, we may introduce so many specious factors that our analysis becomes difficult, complex, and unintelligible. If we scope our system too small, we may miss a factor that is significant to the results. To complicate matters further, we must consider interaction effects as well as the size of the system itself. It is the system definition that can make system testing so difficult.

6.25.1.2 Defining Too Tightly Leads to an Illusion of a System

An extremely "tight" definition of a system can lead to some illusions. For example, we know from the first law of thermodynamics that the change in energy of a close

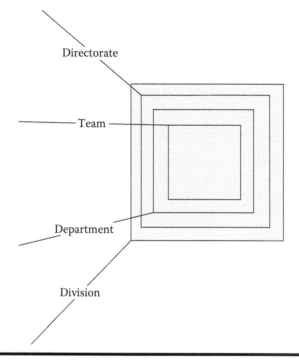

Figure 6.11 The system definition problem.

system is equal to zero. However, for practical purposes, we rarely see any so-called system that could be considered truly closed. In fact, the more common open systems are what can lead to dissipative functions and the attendant bizarre behaviors seen in these systems.

One helpful aid to this definitional issue is to diagram the system, considering all inputs and outputs and relations. While this technique is highly qualitative, we think it helps clarify our definition and provides a target for discussion for those who disagree. Additionally, if we have a hierarchy of systems as in Figure 6.11, we can use our work if our initial stab at system definition does not yield the results we expected.

6.25.2 Outside System

6.25.2.1 Again, Definition of System Limits Implicit

Defining what is outside the system has all the same issues as defining what is inside the system because we are looking at the same problem. Once again, we suggest creating a graphical definition of the system and using some common sense. We can always adjust our definition later if our initial choice is ineffective.

6.25.2.2 What Definition of System Makes Sense?

In general, we want to capture the factors, both control and noise factors, that have a measurable effect on the behavior of what we might call the *system*. We might have to use screening experiments to help us make this choice, as screening experiments help to elicit at least the main factors leading to observed behaviors.

6.25.2.3 Are We Eliminating Inputs?

We generally do not eliminate inputs in a test case. We may model some of the inputs (factors) as noise, but if they are naturally part of what we would expect the complex and embedded product to experience, then we leave these inputs in the test case. We should have established which factors are significant already.

6.25.3 Conclusion

To some extent, system definitions are nearly always arbitrary. We can see this situation when we deal with open and closed systems concepts. If we make the system large enough, we are talking about a closed system; if we make our system small enough, we can expect to see open system phenomena. At a minimum, we should define our system boundaries because we may have some implicit assumptions tied to these boundaries.

6.26 Ambiguity/Clarity

6.26.1 Ambiguity

6.26.1.1 Can Be More Realistic

We find ambiguity of interpretation every day because not all of us "see" the same thing. Hence, a test case yielding an ambiguous result is not entirely unexpected. In some cases, the ambiguity exists because we do not know enough about the system to eliminate the ambiguity. In other cases, we may have an experimental factor that lies somewhere between truly significant and insignificant—it has just enough influence on the results to tip our measurements one way or the other.

6.26.1.2 Cascading Failure with Unknown "Root" Cause

In an ambiguous situation, we may see cascading failures occurring so quickly that we are unable to assign a root cause to the failure mode. We can expect to see this kind of occurrence when we have massively catastrophic failures where the initiating point is not detected by our measuring equipment.

6.26.1.3 Buildup or Precipitating Event

We can also see failures that have twofold root cause mechanisms composed of a buildup of a condition followed by a precipitating event. It is not uncommon to see naïve engineers assign the root cause to the precipitating event and then fail to understand why the precipitating event does not always cause the failure to occur.

6.26.2 Clarity

6.26.2.1 Usually from One Factor at a Time

One factor at a time experiments/tests can yield the illusion of clarity. We need to consider that one factor at a time is rarely seen in the field. We might even take some time when analyzing results to ensure that the clarity we think we are seeing is better than an illusion driven by the measurement and test setup techniques we are using.

6.26.2.2 Can Introduce Distortion as We Force Situation to Be Clear

Our attempts to clarify a naturally ambiguous situation may lead to test execution that imposes clarity where none truly exists. These distorted results can lead to distorted decisions.

6.26.3 Conclusion

We recommend that the test engineer roll with the ambiguity rather than enforce an artificial clarity onto the results. If a rational method for introducing clearer results exists, fine; however, the imposition of false clarity will not make the results more meaningful.

6.27 Sensitive/Insensitive

We might consider controlling sensitivity depending on the level of symptom that we are interested in discerning. Additionally, if we have an extremely sensitive system, we are likely to see more noise in the data. By filtering the system with either hardware or software, we are effectively desensitizing the system; that is, we are trying to estimate the signal in the system.

6.27.1 Sensitive

6.27.1.1 Increase in Sensitivity May Increase Noise without Increasing Information

As we mentioned in the first part of this section, an increase in sensitivity has the trade-off of increased noise in the measurement. We would expect to see some data jitter if we were to plot the raw data. If all we are doing is increasing the sensitivity of the measuring device without a corresponding increase in meaningful information, then we are wasting our resources.

Signal detection can be difficult when the signal is very weak (not necessarily an electronic signal!). If detection is difficult, estimation will most likely be even more difficult than that because the filtering algorithm may identify the signal as part of the noise. If the signal is coherent and predictable, however, we can enhance our detection by estimating the next point.

6.27.1.2 Must Be Expressed

We should always maintain a measurement uncertainty budget for each measuring device we use. We need to recognize that all measurements are wrong in that the measured value and the "true" value are always somewhat different. The difference between them is the measurement error, which is meant in the sense of a natural discrepancy instead of erroneous data collection. This true value is never precisely known, nor is the measurement error. We know that the measure value, T, is a combination of the true value and the uncertainty for measuring T.

$$T = T_m + U_t \tag{1}$$

We can have uncertainties introduced by both calibration and resolution (discrimination). We expect to see a probability distribution around a central tendency (often the so-called normal distribution). We might see such components as the following:

- Calibration
- The measuring instrument itself
- The measurement
- Any fixtures or other connections
- Environment
- Method of analysis
- Sample size (how many measurements we took)

6.27.1.3 What Are We Measuring?

The title of this section is not silly. One of the first questions we need to ask ourselves is what, exactly, are we measuring and why are we measuring it? In many cases, our measurement system/process will be adequate for the task. We then expect to see a very small quantity of our tolerance being consumed by gage error (the rule of thumb is 10% or 0.1). When we are measuring 200 V/m in field intensity measurements, we do not worry too much about millivolts.

In some cases, we may be in the vicinity of a boundary; for example, a voltage limit. If the boundary is considered critical, we may choose to use a sensitive measurement to assess the possibility of crossing the boundary. We know some cases during voltage decay tests where the oscillation of the voltage around the boundary is sufficient to cause a microcontroller to latch up (cease to function). When this latch-up happens, we are less interested in the precise measurement of the oscillation than we are in the lack of deterministic protection for the circuit.

6.27.2 Insensitive

6.27.2.1 Noise May Be Meaningful

In some cases, the presence of noise may actually be indicative of an aperiodic signal. An aperiodic signal may have meaning depending on the effects that arise from interpretation of this signal by our equipment.

In other cases, injections of small amounts of noise may actually lead to improved measurement by improving the sensitivity of the measurement. That is, if we know what we are injecting, we can estimate the value of the signal we are trying to measure.

6.27.2.2 Can Lead to Data Striation (Lack of Discrimination)

If the measuring device is insufficiently precise (related to data variance), then we will most likely see striated data. These data will occupy a small set of discrete values and have a typical "banded" appearance when plotted as an individual-value dot plot (the "stacked BB" plot). Our measuring equipment does not have the ability to discern values between the observed discrete values. Sometimes, this fact may not be significant to the final result. However, if we are designing production automated test equipment that must detect if a gauge pointer is outside the tolerance limits, then we may need to improve the precision of the instrument.

6.27.2.3 May Fail to Detect a True Failure

Coarse measurements may lead to a situation where we completely miss a failure because we are not detecting it. For example, if we design production automated test equipment that only determines if an LCD display illuminates and shows a few patterns, we will not be ascertaining if the text messages shown on that display are correct. Basically, we would be "blind" to the possibility that the messaging system is malfunctioning.

6.27.3 Conclusion

In general, increased sensitivity is probably less risky than decreased sensitivity. We will most likely want to use statistical analyses (e.g., control charts) to separate assignable causes from random variation in the measurement system. Additionally, with increased sensitivity, we are less likely to completely overlook a failure mode. These considerations will be balanced with the cost of the measuring instrumentation.

6.28 Long-Range/Short-Range

6.28.1 Short-Range

6.28.1.1 Can Fail to Detect Time-Dependent Defects

One of the most difficult failures to detect is the one that occurs in a timed test after we have stopped testing! We know of one case where the failure would occur only if another device in the system (from another supplier) ran for a certain number of hours. The tester has a dilemma in this case, because the question then becomes, "How long is enough?"

One potential solution to the short-range problem (i.e., we cannot test forever) with software in particular is to examine the storage locations that function as accumulators. Typical accumulators might be odometers, hour meters, kilowatt-hours, and similar items. These accumulators are effectively integrating instantaneous values. Most storage locations for software will be addressable by bytes or words or some other unit of measurement. We usually run into problems with our software accumulators when we shift from one byte or word to another byte or word; that is, we have boundary issues at the byte or word boundaries. The goal of the tester, then, should be to inject a value using a tool just short of the boundary and let the accumulator then gather more values until it passes through the byte or word boundary. With software storage, the most important boundary is the limiting byte or word that basically tells us our storage is full. With embedded software, it is not too difficult to overflow such a boundary, leading to potential overwrites of critical areas in storage.

6.28.1.2 Less Expensive, Get "Results" More Quickly

Short-range verifications/validations take less time and therefore are less likely to incur the costs associated with long-range testing that consumes testing devices and fixtures as well as the required staffing. As always with testing, we need to balance the risk against the benefit of the quicker timing. Short-range testing may be completely adequate if we are dealing with deterministic items that are not particularly time dependent.

We can also treat the short-range testing as if it were a qualitative test and use it to make a decision as to whether we proceed with long-range testing. If the short-range testing already yields a substantial number of defects on the part of the product, we may decide it is too expensive to run the long-range testing until the design team cleans up the product. When we use this approach, we are deliberately using the short-range testing as a tool to tell us how to proceed with the rest of our test plan. The approach should be documented as part of the verification/validation process, particularly when the customer is involved in the testing or in test approvals.

6.28.2 Long-Range

Long-range testing has two primary purposes:

- Detection of durability problems with the product (usually related to hardware)
- Detection of software boundary problems with accumulators

Because of the increased cost of the long-range testing, we should know the types of failure modes that we are seeking. As with all tests, we would state the *purpose* of the test in our test plan.

6.28.2.1 Must Start Early to Obtain Results before Launch

Long-range testing will often take substantially more planning than short-range testing. We need to ensure that test facilities (chambers, fixtures, instruments) are available for the duration of the test. We need to also make available or hire the staff needed to run such a test. In the automotive industry, it is not uncommon to see durability/reliability tests that run for several months.

6.28.2.2 Costly

Long-range tests are costly by nature. We can improve on the cost by installing automatic devices that will record events and stop the test when an anomaly occurs. Unfortunately, we are not always able to design such a control, but they eliminate the need for human eyes when we can use them. The control devices do not, however, eliminate the need for instrumentation, chambers, and the other paraphernalia needed for the long-range testing.

We might ask ourselves if our long-range approach is unintelligent. We can sometimes use an accelerated modality without increasing the risk, or we can design the test to specifically target the types of failures we expect to see in time-dependen4ptt failure modes.

6.28.3 Conclusion

In general, we would like to see the test team be absolutely certain about what they are trying to find. Short-range testing can reveal a subset of failure modes, and long-range testing will tend to reveal time-dependent failure modes. As we indicated, software that uses accumulators (storage used to integrate instantaneous values) is often vulnerable to time-dependent failure modes.

We execute long-range testing in an attempt to account for latent defects that only appear after substantial amounts of test time. As with any long-term test, we must remain cognizant of the cost of the test; for example, a power module test may require expensive loads, high power (high wattage, high voltage, high current) supplies, and special harnesses. Short-range tests comprise what we usually conduct under customer requirements.

6.29 Costly/Cheap

6.29.1 Cheap (Less Expensive)

6.29.1.1 Qualitative Testing If Used as a Pointer

One of the characteristics of qualitative testing, in addition to speed, is that it usually costs much less than quantitative testing. We can use this characteristic to determine the necessity of the more expensive testing. For example, reliability demonstration testing can easily run into tens of thousands of dollars. However, if we have field data for an existing product and we test a small sample of this older product at the same time we test the new product, then we can save both time and money. If the new product is not demonstrably better than the old product, we may need to make a business decision to improve the new product.

6.29.1.2 May Miss Failure Modes

If our measurement system has poor resolution, we may miss some failure modes because we were unable to detect them. For example, if we have very old oscilloscopes and we challenge our product with some high-speed transients, the device will not be able to represent the transient signal accurately. When we are looking at expenses for a test, we need to consider the types of failure modes that are possible. One well-known way to reflect on failure modes is the DFMEA, which explicitly attempts to assess potential failure modes. We can use the DFMEA to drive testing and control our costs to what seems meaningful.

6.29.1.3 Purpose of Test Must Be Explicit

We should always state the purpose for each test. Doing this simple act eliminates tests that serve no useful function. For example, commercial vehicle manufacturers often require their suppliers to perform salt spray tests, even when it is unlikely that components in the cabin of the vehicle will ever see salt water, salt spray, or salt fog. Most of the time, performing these tests makes little sense.

6.29.2 Costly (Expensive)

6.29.2.1 Significant Qualitative and Quantitative Testing

When qualitative and semi-quantitative testing indicates that further testing should occur, we can proceed economically. For example, although we may be moving into the area of expensive testing, we can diminish the costs by developing our own fixtures. When we are an electronics supplier, we can use our own internal talent to build custom devices for testing rather than buying expensive testers from vendors. We know that "buy, not build" decisions are often cheaper; however, when we need a custom tester, it is less expensive to build it ourselves than to try and modify an existing test measuring device.

6.29.2.2 Target Is to Yield Meaningful Failure Modes

Even though some of our testing may be expensive, we can diminish the costs by looking at only meaningful failure modes. Here, again, the DFMEA can help assess the severity of the failure modes and the probability of the untoward event actually occurring. The DFMEA should provide us with a rationale for choosing one test over another.

6.29.2.3 What Is the Concatenation of Worst-Case Environments?

If we concatenate our worst-case environments to produce a multi-environmental stress scenario, we can also save some expense. An example of this approach occurs when we combine vibration, humidity, and temperature testing simultaneously.

6.29.3 Conclusion

We must balance the *potential* cost of *not* performing a test (lawsuits, field returns, zero kilometer returns, returned materials authorizations [RMA]) against the cost of performing the test. Potential costs are difficult to assign and define—we sometimes can get help from historical data, but often we have no data to support our assumptions.

6.30 Flexible/Inflexible

6.30.1 Inflexible

6.30.1.1 Do Not Change a Designed Experiment Midstream

Changing a designed experiment in midstream is generally *not* a good idea. In the best of cases, we might be able to handle the results as a split experiment. In the worst of cases, we would have to start all over again. Designed experiments are just that: designed. We gain tremendous efficiency and power from following the recipe exactly.

6.30.1.2 Do Not Allow Rules to Blind

While we do not want to change designed experiments in the middle of the treatments, for other kinds of testing, we want to proceed with an nonjudgmental awareness. Policies and procedures will sometimes overstabilize a system to the point where they become ingrained in the test team, who will no longer think about these actions and reflect on whether they still have a purpose. This situation is a form of bigotry against change and should be avoided for obvious reasons.

6.30.1.3 Consider Alternative Testing Techniques

When one of us first began reliability testing per se, we used the reliability demonstration approach, which required huge amounts of sample pieces and much time and expense. Sometimes, we still must set up our reliability tests this way due to customer requirements. However, when we are in the position of having legacy parts, especially if they are similar to a new design, we can use the fact that we have field data on the older parts and perform a baseline versus comparison (B vs. C) test with significantly fewer test units. The statistics will usually be something like Tukey's quick test, which can tell us something about the difference or similarity (reject or not reject the null hypothesis about the mean) in means with six parts of the old material and six parts of the new material. We call this a semi-quantitative or qualitative test because we are really trying to discern if the new parts are better than the old ones. Presumably the old parts were good enough for somebody, so they represent some level of "acceptable" field performance. We would like our new parts to be better than the old parts and a good B vs. C test can give us that information. Indeed, we use this test approach particularly with highly accelerated life tests such as thermal shock, mechanical shock, and resonant frequency testing.

6.30.2 Flexible

6.30.2.1 May Bend the Rules Such That Results Are Meaningless

If our testing approach completely lacks a system, we will not have meaningful results. For example, we do not want to change a designed experiment midstream, precisely because the results may become utterly meaningless. On the other hand, we have also conducted electromagnetic compliance (EMC) tests with the antenna pushed closer to the sample than the standards allow. We are aware of the so-called near-field effects; however, this move allowed us to reach 200 V/m field intensity over most of the range. If the product did not pass this test, it did not make any sense to send the parts out to a lab that had the capability to achieve 200 V/m with the correct distance and expensive amplifier.

6.30.2.2 Requires Careful Documentation

If we are doing exploratory testing, especially with software, we will want to record the steps we take as we allow the tester to "play" with the unit under test. Sometimes the test engineer will come up with a creative or extremely destructive test that we want to add to our test suite. Good documentation allows us to do that. Over time, we expect our test suites to grow as we learn more about the product using exploratory testing.

6.30.2.3 Not Appropriate for Designed Experiments

As we have mentioned, we deprecate modifications to designed experiments. Recovery often involves more work than it is worth. Sometimes the best option is to simply stop the experiment if a noise value cannot be blocked out or some other extraneous input is butchering our results.

6.30.3 Conclusion

We want flexibility insofar as we are allowed to design our test suites and experiments. Once we have committed to a test approach, however, excessive flexibility will only raise the cost of the testing. In short, we would expect to see a rigorous business case before changing our test plans significantly.

6.31 Parameter/Tolerance

Parameter and tolerance design were recommended in that order in most of the works by Genichi Taguchi. We believe the approach is sound based on our experience.

6.31.1 Parameter Approach

All Taguchi approaches rely heavily on his method of experimental design. It is not necessary, however, to use the Taguchi experimental design approach if we have the tools to support the classic approach or the response surface methodologies.

6.31.1.1 Robust Design

The goal is to achieve a robust design. A design is robust when it is immune to extraneous influences, some of which may be considered noise. In essence, we are trying to desensitize the parameters of the product such that it will behave appropriately regardless of the circumstance. Obviously, some limits on these parameters will exist because of cost concerns. Nonetheless, the robust design approach should make for a more reliable product.

We use the designed experiment approach to assess where we need to improve the design to make it more robust. If a parameter is extremely sensitive, we can either use it for control of the product or make it less sensitive. Many products that support DOE can also support the response surface approach and show contour diagrams that help the user identify regions that vary little as external values change significantly. These regions are areas of robustness for the product design.

We must be careful when running designed experiments. We may be able to optimize the product *performance* while at the same time degrading the product *robustness*. In some situations, we will find that robustness and performance are competing factors. The test team will provide the information to the design team. The design team will make the appropriate decision with customer input.

6.31.1.2 Use Derating

We can increase robustness by derating parts. We derate a part by using design limits that are *inside* the real design limits. In essence, we are giving ourselves a safety margin on every component. The derating approach can become expensive if used with abandon. On the other hand, the approach is simple to implement and will nearly always provide a boost to the robustness (reliability) of the final product.

6.31.1.3 Lowest Cost

Parameter design is less costly than tolerance design. At this stage, we are dealing with attributes and variables that do not require tolerances.

6.31.2 Tolerance Approach

6.31.2.1 Expensive

The tolerance approach is expensive because the supplier will now have to meet tolerance requirements. Furthermore, the designers will have to verify that the stack-ups (root sum of the square, Monte Carlo, or both approaches) are within design guidelines. Designed experiments can be used during tolerance design just as they were during parameter design.

If we are an automotive firm (supplier or customer), we will require our suppliers to meet the requirements of the product part approval process (PPAP) to make an appropriate submission. The PPAP allows suppliers to check their work and product one last time before submitting supporting evidence to their customer. A PPAP has 18 components:

1. Design records
2. Authorized engineering change documentation
3. Customer engineering approval
4. Design failure mode and effects analyses (DFMEA)
5. Process flow diagrams
6. Process failure mode and effects analyses (PFMEA)
7. Control plan
8. Measurement system analysis studies
9. Dimensional results (tolerances!)
10. Record of material/performance test results (lab results)
11. Initial process studies
12. Qualified laboratory documentation
13. Appearance approval report (AAR)

14. Sample production parts
15. Master sample
16. Checking aids
17. Customer-specific requirements
18. Part submission warrant (PSW)

The PPAP documentation is extensive but it helps to provide protection for both supplier and customer. When a customer approves a PPAP document, it becomes a form of acceptance authorization.

6.31.2.2 Use Only When Necessary

Because of the expensive nature of tolerance requirements, we should be wary of including tolerance requirements for a host of dimensions. The design experiments should tell us which dimensions are significant to performance, and these are the ones for which we would then use tolerances.

6.31.2.3 DOE for Both

As we have seen, both parameter and tolerance design make use of DOE. We strongly recommend the use of designed experiments because it is one of the most efficient methods for simultaneous manipulation of causative factors. The DOE approach will also tell us which factors are most significant to the quality of the product.

6.31.3 Conclusion

The bottom line is to always consider the bottom line. We should fix what we can fix during the less expensive portion of the design process and save only those items that absolutely need tight tolerances for later. It does not make sense to do the expensive portion first and the less costly portion afterward.

We expect both parameter and tolerance testing to occur during the preliminary design phases, especially if we have simulation capability. The goal is to achieve a robust product as much as possible with parameter design and only proceed to do tolerances if we must do so. Tolerance specifications increase the cost of supplied parts, so we want to keep tolerances to what is needed and no more.

6.32 Standard/Failure

6.32.1 Test to Standard

Many customers will ask a laboratory to test to specific standards—this allows them to compare apples to apples among suppliers.

6.32.1.1 Repeatable, Reproducible

Standards help ensure some level of repeatability and, to some extent, reproducibility. We get repeatability because the tests will be defined to be the same among suppliers. We get some reproducibility because the procedure is often well defined, which helps eliminate some of the variation.

We know, however, that repeatability is difficult with electromagnetic compatibility in particular. A small shift of the sample or the antenna can produce significantly different results. Yet these are standardized tests (U.S. DoD, U.K. MoD, SAE, etc.). So, in some cases, the repeatability and reproducibility can be illusory.

6.32.1.2 May Not Induce Failure

Some of the standardized tests are weak enough that we may never see a failure. One example we have seen occurred when the customer supplied accelerometer recordings from actual truck drives. Unfortunately, the routes chosen were not particularly challenging, and this "realistic" test never yielded any failures. In this situation, the so-called realistic test may not have been what the customer and the supplier needed to properly characterize the failure modes of the product.

Many of the standards are quiet about sample sizes. If the probability of failure is below the threshold of the unit count we are using, we will only see a failure by happy circumstance.

We are somewhat leery of standardized tests because we have marginal products released to market after passing standardized tests. We think we have a better approach to characterizing the units under test.

6.32.1.3 Where Is the Failure, Then? We Don't Know!

The previous sections should have made it clear that even after passing standardized tests, we may have latent failure modes that will only show up in the field! Even when we consider an observed failure to be unlikely to occur in the field, we have still discovered a real failure mode. We can report extreme failures as observations or incidents to document their presence and, if we are suppliers, negotiate with the customer as to what priority to assign this new failure. If the failure is truly off the wall, the customer may log it as an observation and forbid further work on it.

6.32.2 Test to Failure

We have at least two levels of test to failure with our test regime:

1. Test until the part no longer functions as required.
2. Test to complete destruction of all or part of the product.

Both approaches have their advantages.

6.32.2.1 May Be Repeatable, Reproducible

We see no reason why test to failure cannot be as repeatable and reproducible as typical standards-based testing. Where we may deviate is if we need to go outside the standard definition of the test in order to induce the failure. We can help ourselves and others by documenting our experimental approach in enough detail that any other test group would most likely see the same kind of failure.

We can also join a standards committee with a commitment to upgrade the standards-based approach so that we are more likely to educe the latent failure modes. The previous sentence is not a silly comment or idea—it represents one of the ways a professional can influence the rest of the profession in profound ways.

6.32.2.2 More Complete Characterization

Test to failure approaches provide more complete characterization than simply "going through the motions" with a standards-based test case or plan. We know we will drive beyond the design limits to the failure. The interesting part of all this occurs when the failure limit is surprisingly close to the design limit. Such a situation should produce heartburn for the designers because it is likely in this situation that we are dealing with a marginal design.

It would also be helpful when we are driving to a failure to have sample product built with some variation in the components so that we can see the result of our harsh test regimes on a product that has some natural variation. Certainly, adding variation is a more realistic approach because this will occur during normal manufacturing anyhow.

Testing to destruction goes beyond testing to failure; at least in most cases it does. When we test to destruction, we have at least one component, software or hardware, completely cease to function—in fact, it ceases to function irreparably. At the destruct limit, we have more completely characterized the product than at the failure limit. The only question at this point would be whether or not we proceed to complete destruction of the unit under test. We believe that this is a judgment call that should be based on the experience of the test team as well as the severity potential failures in the product. The failure of an anti-locking brake system (ABS) is presumably more severe than the failure of the in-dash radio in a motor vehicle.

6.32.2.3 Test Until Failure Modes Are Induced

We always push to test until we see some failures. If we do not push to the failure limit we will not know where that limit really is, and this situation can result in selling a marginal or dangerous product. We consider testing to failure to be the *minimum* we should do, yet that is more than we have seen most test groups do when they execute a test suite.

6.32.2.4 Customer May Panic

Testing to failure shows the customer the part can, in fact, fail. This happens, of course, but the information may not be well received by the hypersensitive customer who expects the product to never fail—ever—and certainly not during any product verification or testing. The benefits of performing testing to the limits of the product outweigh the transient response of the customer who spontaneously erupts when a failure is reported. It is better to understand the limitations of the product early—even before the product gets shipped to the customer.

6.32.3 Conclusion

The risks associated with testing to failure are small compared to the risk of not knowing the product's limits and failure characteristics. It may be a little bit uncomfortable to have these early failure discussions with the customer. These product failure conversations become exponentially more difficult once the product is shipped to the world.

6.32.4 Small Sample/Large Sample: Small Sample

6.32.4.1 Enough for Statistical Analysis?

We are always in a battle between meaningful sample sizes and cost. If the part is particularly expensive to build, we may receive pressure from management to find techniques that allow us to use smaller sample sizes. This situation is particularly true in the case of reliability testing, where we often have requirements for 90% reliability at 90% confidence—we need 20 pieces per test to achieve the appropriate level of reliability demonstration.

On the other hand, if we are truly looking at probabilistic failure modes, the failure mode will show up in a small sample size only by happenstance. We can use the DFMEA to help us determine the severity or criticality of the issue. That means we *must* see to it that the DFMEA documentation and the test plan correlate with each other. We do not have to stop at only the items identified in the DFMEA, but we should at least meet halfway with the DFMEA.

6.32.4.2 t-Test and Distribution

The t-test is a common test for small sample sizes. It does, however, have some caveats; for example,

- Requires reasonable normality of the sample means
- Requires that distribution be close to normal for small sample sizes

■ Must have homoscedastic population (i.e., we expect our samples to have approximately the same variance)
■ Cannot be used with count data, only continuous, variable data

The nonparametric equivalent is the Mann-Whitney-(Wilcoxon) U test.

6.32.4.3 How to Catch a Failure

We will capture a relatively rare event with small samples only by happenstance. Usually we do not know the probability of a rare failure until the product has been out in the field. Just because we are not observing failures does not mean we do not have latent defects. We may be in a situation where our sample size is so small that we will *never* see this particular failure until a substantial quantity of product has been sold.

One approach to solving this problem is to increase the sample size until we do see failures. This method can become quite expensive if we are unlikely enough to keep missing the guilty parts. We should also revisit the DFMEA to see if we have any high-severity or high-criticality failures we anticipated seeing but have not seen as yet.

6.32.5 Small Sample/Large Sample: Large Sample

6.32.5.1 Increased Likelihood of Detecting a Failure

Thanks to the law of large numbers, we can expect our sample approximations of population parameters to begin to converge on the population values as our sample size increases. The only way we would not see this event occur would be if we were extremely unfortunate in our sample choice or if the data were highly heteroscedastic (variances not equivalent). The trick during testing, then, is to pick *representative* sample sizes that will reveal the defects that we are supposed to discern.

6.32.5.2 Appropriate Distribution

Rest assured that the normal distribution is not always the normal distribution. We rarely see the normal distribution in reliability distribution analyses, where it is much more likely that we will see the following probability distribution functions (at least in approximation):

■ Weibull
■ Log-normal
■ Gamma
■ Exponential
■ Pseudo-normal

We should also realize that these *continuous* distributions are approximations based on discrete samples. In some cases, it makes more sense to use discrete probability distribution functions to analyze our results. For example, we can model a relatively rare occurrence as a defect arrival rate using the Poisson process.

6.32.5.3 Confidence Level

A confidence level provides some information about the probability that a calculated statistic or parameter falls between two limits. The complement of the confidence is the significance; that is, if our confidence probability is 0.95 (95%), then our significance is 0.05 (5%). Counterintuitively, the higher the confidence, the wider the limits; if we think about this, it becomes obvious that wider limits are more likely to contain the true value.

6.32.6 Conclusion

We always recommend testing to failure, if not destruction, of the product. We believe that full characterization is the best way to proceed for the following reasons:

- If the customer claims to be seeing failure modes that we know can only be generated by abuse outside the design limits, we have a counter-case.
- We know how "close" the failure limit is to the design limit.
- We understand our product more completely.

6.33 "Good" "Bad" Testing

"Bad" testing is not so much bad per se as it is suboptimal. Particularly in automotive testing, we seem to waste a lot of time going through the motions rather than *attacking* our products and squeezing every bit of characterization we can out of the product.

6.33.1 Bad

6.33.1.1 "Bad" Testing Tells Us Nothing New

Perhaps the late Edsger Dijkstra made the single most famous comment about software testing: "Program testing can be used to show the presence of bugs, but never to show their absence!" This statement is also true for hardware testing and is related to the logical difficulties of reasoning from observed effects to causes (also know as *affirming the consequent*, whose corollary is *denying the antecedent*).

In the automotive industry, it is common to conduct design testing called *design verification testing* or DVT. Once we have production parts from a production process, we are supposed to run *production validation testing* or PVT. Often PVT is an iteration of DVT. If large parts of the product have not changed, we see no reason to conduct these tests again unless we think we will see something different in the results than what we saw the first time. A test that does not inform us is a waste of time and resources.

The only reason for conducting iterative PVT is that this testing fouses on the influences specific to the production processes, while the DVT parts may have been built via prototype processes as well as whatever production processes may have been in place at early stages of product development. When DVT parts are from the full production system, the PVT becomes redundant and of little value because it tells us nothing new about our product.

Every test case we execute should be performed for the purpose of learning about our product—software or hardware. Going through the motions wastes everybody's time. Passing a suite of test cases does not really tell us much of anything—we are effectively in the scenario described by Dijkstra's quote.

6.33.1.2 "Bad" Testing Is Artificial

Artificial testing occurs because we isolate the units under test from extraneous influences; for example, with electromagnetic compatibility testing, we use a shielded chamber to ensure that we are reading the true values for emissions and subjecting the unit to only the antenna signal for susceptibility. However, actual product units must face a myriad of conditions in reality. That is why we had a flurry of activity with stirred-mode EMC testing a decade or so ago—the practitioners were attempting to increase the verisimilitude of the test. Unfortunately, stirred mode was generally unrepeatable and thus considered unacceptable by the testing community at large.

We have both worked for companies related to the commercial vehicle (trucking) business. In many cases, we have seen subsystems pass bench testing with flying colors and fail within minutes after being installed on a real vehicle. One issue is the installation itself and another is the random intrusion of noise. Noise is not always easy to simulate, particularly incidents that occur truly aperiodically. Somehow the other subsystems never seem to function as well as the neat, little environment on the bench. Batteries show sagging output, impedances are different because of harness routings, and bus timings change.

Another component of this artificial approach to testing is the soft concept of "attitude." Our tests must always be looking for anomalies and driving to failure, not trying to get the product to pass. We have seen good software testers become ineffective at testing when they tried to become software engineers—for some reason they were too sympathetic with the developers and began to try and get the software to pass.

6.33.1.3 "Bad" Testing Is Inefficient

Many times, the testing required by customers will reflect an OFAT approach to testing. This approach, in itself, is a liability because we are unlikely to see factor interactions except by accident. Furthermore, OFAT testing maximizes test case quantity while minimizing the information derived from the testing.

Repeated testing with the same or similar results is generally a waste of time unless we are trying to show that a change to the product did not cause a change to the test results. Unfortunately, this path also has problems: if we passed the first time (and really learned nothing) and we passed the second, we still learned nothing about the product. As we mention elsewhere, we need appropriate sample sizes and test to failure (or destruction) to really characterize a product, particularly after a change.

6.33.2 Confirm Fixes from Previous Testing

When dealing with software issues, we sometimes see an error creep from one revision to another without correction. The literature may call this approach defect containment. Poor defect containment can represent some serious issues:

- A terrible relationship between testing and development
- Ineptitude on the part of the developers
- Deplorable software development management

Furthermore, the reappearance of previously identified defects puts more stress on the testing team.

One method to help manage this situation occurs when our test teams attack the code areas where they know changes should have occurred. Alternatively, they can request evidence of appropriate unit testing. Even so, we would recommend the attack approach to keep the process clean. Each release of software should be known to be clean of all previously identified anomalies, thus making it easier to determine if new code introduced any previously unseen defects.

6.33.3 Good

6.33.3.1 "Good" Testing Tells Us Something New

Testing that tells us something new about the product is a good alternative to go-through-the-motions testing to fill in a box on a customer document. Each test case should be eliciting more information from the product—in essence, persuading the product to speak to us. It is not uncommon during automotive testing to execute a suite of tests called design verification and then follow that later on with the same suite of tests and call it production validation. The putative purpose of the production validation is to verify that the manufacturing lines are producing the product correctly by performing design-type tests. This approach makes little or no sense, not to mention the fact that the sample sizes are usually so small it is unlikely that a probabilistic issue will be captured—except, perhaps, by blind luck.

6.33.3.2 "Good" Testing Stimulates Multiple Responses

Any time we can make use of a designed experiment, we should do so. In some cases, we may be using a sequential simplex approach, which is more of a testing search algorithm. Both methods can be highly efficient, looking at multiple factors simultaneously as well as multiple responses. If we are recording multiple responses, optimization may become difficult because we may have competing responses; for example, we may wish to select a good fertilizer but we also want to optimize for cost. This situation simply makes the optimization problem somewhat more complicated.

6.33.3.3 "Good" Testing Produces a Robust Product

In essence, robust testing leads to a robust product. We design our tests to verify that the product is insensitive to extraneous influences where we want it to be insensitive. We may use more sensitive factors as control factors. The point here is that we do not want to see the design collapsing under the influence of environment noise, wear-out, or user ineptitude.

With hardware, we can use response surface methodologies, a sophisticated designed experiment approach, to identify areas of sensitivity and insensitivity. Once we know which factors to analyze, we can use an optimization technique like particle swarm optimization or a genetic algorithm to put the design in the region where we want it. In fact, the ability to use an optimization algorithm is one of the prime reasons to use the response surface approach in the first place. Additionally, a response surface analysis can handle nonlinear responses gracefully, providing a more accurate model than more linear approaches.

6.33.4 Conclusion

Obviously, we favor "good" testing because it helps us release a high-quality product.

6.34 Parallel/Sequential

6.34.1 Parallel

6.34.1.1 Increased Throughput

Parallel testing is much like concurrent engineering—we take tests that have no dependencies on other tests and no major person or equipment constraints and we run them simultaneously to increase our throughput. This approach is particularly easy to implement with typical OFAT testing and with standards-based testing.

6.34.1.2 May Need More Samples Than Necessary

With parallel testing, we need enough samples to be able to run the test simultaneously, whereas with sequential testing, we can use the samples from previous tests. Some customers put such an order into their test requirements. Unfortunately, we still nearly always run a OFAT test even though it is sequential.

6.34.2 Sequential

6.34.2.1 Specified by Abraham Wald in World War II

The great statistician, Abraham Wald, developed the sequential testing statistical approach during World War II. The method was considered so important to the war effort that he was not allowed to publish his work until after 1947. The method allows for performing one test after another (sequential) and updating one's statistics with the completion of each test. The most famous instantiation is the sequential probability ratio test (SPRT).

6.34.2.1.1 What are fail, don't know, and pass zones?

We reference Figure 6.12 for this description. We execute the tests in the following way:

1. Calculate meaningful boundaries.
2. Calculate a termination point for both fails and successes.
3. For every success we plot a line one unit to the right.
4. For every fail we plot a line one unit up.
5. We expect to see the tests zigzag through the test region.

A fail zone is effectively the reject region. A pass zone is effectively the success region. The "don't know" region occurs above or to the right of the termination lines (they

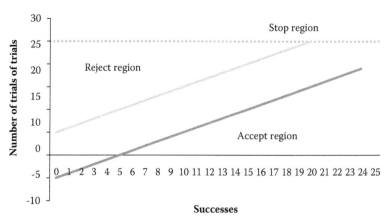

Figure 6.12 Sequential probability ratio test (SPRT) graph.

make the "point" on the tipped cylinder in the graph). This test is often used in medical studies because it is reasonably humane when the termination values are set fairly low.

6.34.2.2 May Take a Long Time to Make a Decision

If the termination values are not calculated correctly, the test may take too long to yield meaningful results, especially if the fails and successes alternate. This alternation will yield a stair-step appearance to the results as they are plotted against the limits.

6.34.2.3 Works Best with Really Good or Really Bad

This test works really well with data where the fails are numerous or the successes are numerous, yield quick results and speedy test termination. One of the other issues with the "meandering" effect is that we consume time. As we consume time, it is possible the experimental conditions will change during the course of execution of test cases.

6.34.2.4 All Three "Zones" Specified Up Front

The three zones should be specified before the test starts. It is out of the scope of this book to explain the statistics of this test; we simply need to be aware that this approach exists and has some good uses, particularly with reliability testing.

6.34.3 Conclusion

The choice between parallel and sequential testing is a true judgment call. Parallel testing allows us to save time in much the same way as concurrent engineering saves design time. On the other hand, we have seen test specifications from customers that provide a test flow with a set of different tests executed sequentially. If a rationale exists for executing tests sequentially, then the approach is logical.

6.35 Fractal/Smooth/Ordered

6.35.1 Fractal Stimulation Is Effectively "Noise"

A fractal will generally have the following features:

- Finely structured at arbitrarily small scales
- Extremely irregular
- Self-similar
- Recursively defined

It is the extreme irregularity that makes the fractal form useful as a noise model. We are not as interested in the fractal graph as we are in the values generated by the function. The irregularity of the fractal boundary can provide a good source of stimulation for the units under test.

6.35.2 Smooth Stimulation Is Not Necessarily "Not-Noise" but Rather Invariant Stimulation

Invariant stimulation is not bad—it is merely another form of stimulation. One example of this kind of usage occurs when we read a test standard and find comments that say the anticipated nominal voltage on a commercial vehicle is roughly 14.2 volts. We can test at that value but it is most likely not particularly realistic.

6.35.2.1 What Are Effects of Long-Term, Invariant Stimulation?

If we choose to go ahead and use invariant stimulation, we need to understand what we are trying to do with the test. For example, it is not uncommon to see invariant high voltages, high humidity, and high temperatures during reliability testing. These values are chosen above the nominal value to achieve a level of test *acceleration* so that the test elapses over some reasonable amount of time—certainly before we launch the product.

6.35.2.2 Can We Detect Self-Stimulated Noise in the Test Unit?

Self-stimulated noise occurs when the nature of the product generates its own noise. An example can occur when we have variation in assembly and the part does not quite vibrate the same way from one piece to another. We have also seen cases where an improperly terminated data bus generated significant emissions noise, affecting other parts in the product—not to mention the effects on nearby subsystems.

Figure 6.13 shows a comparison of the three modalities we have been discussing. Note that, in a general sense, the fractal signal is ultimately ergodic, as it tracks with the sine wave; however, locally it shows a substantial amount of variation.

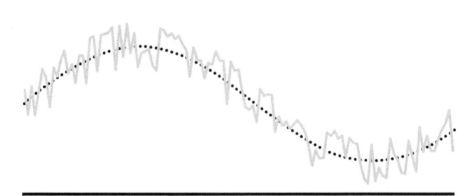

Figure 6.13   Fractal, smooth, and ordered stimulus modalities.

6.35.3 Ordered Stimulation Is Definitely "Not-Noise"

Ordered stimulation or periodic exercising of the stimulus is yet another option that is not bad. We may want to design a test to specifically "hammer" the sample parts. One of the destructive tests we regularly use is the swept sine test on the vibration table. The purpose of this test is to determine the resonant frequencies of the product. Once we have resonant frequency for the product, we can do highly accelerated testing at or near the resonant point. The signal is highly ordered (and smooth) and provides a significant challenge to the parts.

6.35.4 Conclusion

We recommend all three modalities, but especially the fractal and ordered approaches. We have seen situations on motor vehicles where a part becomes partially detached from a mounting and we have a quasi-ordered mechanical stress scenario. On the other hand, we have also seen situations where a fractal approach makes more sense; for example, ambient electromagnetic signals may appear more fractal (although they may not have the self-contained replication factor).

Notes

1. Gilles Deleuze and FÈliz Guatarri, *A Thousand Plateaus: Capitalism and Schizophrenia* (Minneapolis: University of Minnesota Press, 1987), pp. 3–25.
2. Atul Gawande, *The Checklist Manifesto: How to Get Things Right* (New York: Metropolitan Books, 2009), p .
3. Christopher Alexander, *The Timeless Way of Building* (Oxford, UK: Oxford University Press, 1979).

Chapter 7

The Use of Noise

Figure 7.1 shows a multitude of factors that can affect the performance of a product. Because these factors are generally not well controlled, they truly become noise factors rather than control factors. We may be able to simulate this noise in the lab and perform some level of control so we can ascertain the potential behavior of the product under specific experimental conditions.

7.1 Realistic

Noise is a natural component of our existence. Even the "ether" is permeated with noise—our radio telescopes can pick up this background noise from interstellar emissions. We can identify specific kinds of noise:

- User mishandling
- Environmental
- Behavioral degradation
- Part-to-part

These four types of noise are common sources of product breakdown, so they make a great deal of intuitive sense. Replicating them in the laboratory environment can sometimes present difficulties; for example, we may not have a large enough sample to see significant part-to-part variation.

7.2 Can Use Taguchi Approach

Figure 7.2 shows the three high-level steps taken during the Taguchi approach to design. We use designed experiments heavily during the parameter and tolerance design phases.

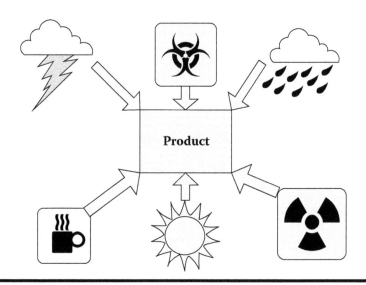

Figure 7.1   Noise and testing.

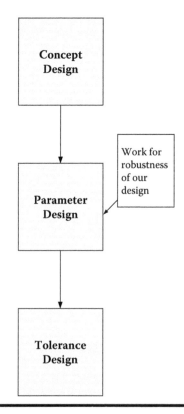

Figure 7.2   Taguchi approach.

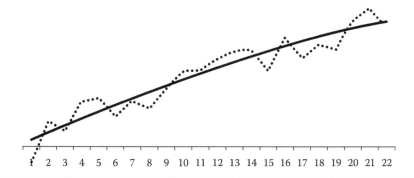

Figure 7.3 **The appearance of noise.**

7.3 Sometimes Difficult to Simulate

True randomness can be difficult to implement and even more difficult to prove. The question we must ask ourselves is how important "true" randomness is to our results. If our real goal is the generation of aperiodic stimuli, then we can do this easily by capturing radio-frequency noise, natural variation in data values, and any other event that is not obviously ergodic in nature.

Sometimes we will want to run tests that use repeatable sequences of pseudo-random values so we have a level of repeatability for future replications. Random number generators provide pseudo-random numbers easily and we can use these during our testing.

7.4 Where to Apply Noise?

We use the term *noise* to indicate an uncontrolled and, sometimes, uncontrollable factor that affects the results of our experimentation (testing). Figure 7.3 shows the presence of noise around the apparent function (whichever function it might be). In short, we have some level of randomization around the theoretical value of our function.

7.5 Exogenous Shocks

An exogenous shock, by definition, is an externally generated, high-intensity event for the product or service. We can classify an exogenous shock as noise, although many practitioners consider noise to be uncontrolled environmental factors constantly applied. The exogenous shock may be an infrequent and aperiodic event—a sort of Murphy's law version of infrequent noise, arriving at the most inopportune moments.

Figure 7.4   Exogenous shocks.

In our simple example, Figure 7.4, our truck is about to run over a deep pothole and is also subject to lightning, severe storms, and hot sun. In the automotive business, especially the subset we call commercial vehicles, it is not uncommon to see the drivers visit every political subdivision in the country, traveling from deserts to mountains to salty ocean-side locations, to snowy roads, and windy plains. All of these environmental incidents can be considered exogenous shocks when sufficiently intense.

Not surprisingly, testing for exogenous shocks often occurs only by happenstance. More enlightened test teams will assault the product with mechanical shock, thermal shock, heavy salt fogs, and severe vibration to determine how the product will perform under severe conditions. We think it is critical that these tests be added to the test suite in any case where the product is even remotely likely to receive exogenous shocks.

7.6 Self-Generated Noise

We need to keep in mind that our product may generate its own noise. With electronics, we often measure this characteristic as conducted or radiated emissions— our product is literally generating a radio frequency (or some other part of the electromagnetic frequency band). If we suspect that our product is making its own noise, it behooves us to perform the appropriate testing to see if this is the case.

Other sources of self-generated noise can include friction between parts.

Notes

1. Gilles Deleuze and Fliz Guatarri, *A Thousand Plateaus: Capitalism and Schizophrenia* (Minneaapolis: University of Minnesota Press, 1987), pp. 3–25.
2. Atul Gawande, *The Checklist Manifesto: How to Get Things Right* (New York: Metropolitan Books, 2009) p.
3. Christopher Alexander, *The Timeless Way of Building* (Oxford, UK: Oxford University Press, 1979).

Chapter 8

How to Perform "Bad" Tests

Bad testing can have a big impact on the customer's experience of the product. This adverse customer impact has an impact on the profit margins of the product and volumes of the product sold, not to mention the impact of potential legal action (see Figure 8.1). Poor testing can be expensive. Spending many hours and much money on testing and missing product nonconformances or other problem areas is worse yet. Not only is this a waste of resources, but it also provides a false sense of security for the organization.

8.1 Do Not

8.1.1 Add Stress

Testing the product to a very low level of stimulus or minimum rigor does not help us learn something from the product.

8.1.2 Go Beyond the Specified "Limits"

Specifications are a good starting point; however, frequently these do not capture the total exposure of stimuli to which a product will be subjected. We have seen many cases where the product passes testing to specification and then summarily fails in the field. Variation of key product design areas as well as variation in the external environment can all add up to produce a failure in the field.

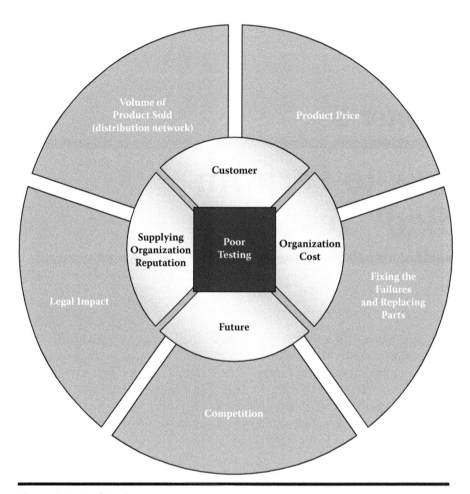

Figure 8.1   Bad testing.

It can take a great amount of time to understand the real demands on the product when it is deployed by the customer. Gathering the information requires instrumentation for measuring sample units while also securing a variety of example uses of the product with sufficient sample size and variation to provide some statistically significant understanding, which takes time. Experience suggests that developmental test engineers rely heavily on standards. These are a starting point, but we always question the percentage of the customer population represented by the standard. One of us has experienced the condition where the product passes the electrical transient portion of a specification only to go on to produce failures in the field. This anomaly is not an infrequent occurrence but all too common—pointing to testing to specifications as a conceivably short-sighted methodology.

8.1.3 Use Unusual Combinations of Events

The real world is more complicated and chaotic than we believe. Testing only to typical combinations is very shortsighted and unrealistic. We can use the Cynefin[1] model to show the sequence of simple to chaotic systems:

- Simple
- Complicated
- Complex
- Chaotic

8.1.4 Check All Inputs and Outputs

We once heard a project manager lament, "If we did the design and development work right, we would not need testing." This is the same mentality that would lead us to test only select inputs and outputs of the product. Development work takes many people and coordination of events as well as a continuous eye for the details. As long as the development includes humans, there will be a need to test with a measure of rigor. Neglecting some of the inputs or outputs means open areas for failures for your customer to find.

8.1.5 Follow Up

Finding a fault or software defect is only the beginning. Once the bug is reported, a follow-up in the form of a corrective action is in order. Will the defect be corrected? In the event that the defect is to be corrected, follow-up testing is necessary. We have seen instances where a fault was reported within the reporting system and the development team indicated that they understood the fault and generated a new part as a result. The test engineer closed the fault report. The new parts were shipped to the field, where they failed forthwith. There was no confirmation (via verification) that the corrective action corrected the defect. Reporting the defect is not the end of the verification work. It is necessary to ensure that corrections address the defects and do not introduce new defects.

8.2 Do

8.2.1 Let the Designers Create the Test Plan

We consider allowing the developers to create the test plan to be a conflict of interest. We know of one egregious case where the designers created an air pressure gauge yet had no tests for air pressure in the test plan. The gauges subsequently leaked and caused a major warranty issue.

The test plan should be created by the independent verification and validation team, whether they are internal or external. Certainly the designers can review the test plan for obvious misunderstanding or misinterpretations, but they are not the authors or final judges of the test plan. We also recommend that the customer be involved in approving the base test plan, although we often leave the customers out of the picture when developing more rigorous product characterization test plans.

8.2.2 Test Only for Nominal Input Values

The automotive test standard SAE J1455 says the nominal voltage for a 12-volt motor vehicle is 14.2 volts. For years, one test group we know tested only at 14.2 volts. It took a minor convulsion of management browbeating to get the test team to switch to random voltages and deliberate slewing of the voltage values. A steady 14.2 volts nominal is not a realistic situation in the field, and we see no reason to test to this value on the bench.

The same approach applies to any other environmental value that can affect system performance. We conduct vibration tests with varying humidity and temperature to add environmental change to the testing. We are considering ways to add electrical transients to this recipe.

8.2.3 Make Test Plans that Are Not Consistent with Historical Evidence

Failure to make plans that are within your means in order to achieve an objective can be as bad or worse than not planning. This blindness to the past is another area of testing failure. Creating a schedule that does not account for the performance and the level of expertise of your organization provides *false hope*. Schedules should be rational and achievable. Every time we crash the test schedule, we take the risk of seeing tired test engineers make mistakes.

8.2.4 Provide a Benevolent, Air-Conditioned Environment

Even in cases where the product will most frequently be used as an air-conditioned environment, we do not consider this the way to test a product to meet a customer's expectation. It is difficult to quantify the total exposure of the product to the various stimuli it will experience during the customer's use.

8.2.5 Forget About Extreme Situations

We use extreme situations to create failures. Sometimes we can also use extreme environments to cause product failure much more quickly than we would find if we were using nominal environmental limits. Even if we warn the product operator/user to avoid extreme environments, we should still know under what conditions the product will fail.

8.2.6 Ignore the FMEA or Fault Tree

Some development organizations do not know about FMEA or fault trees and how these tools benefit the development effort. Some organizations use these tools but use them after the design is completed and ready to go to production—each approach provides similar results. Not knowing about or poor use of these tools provides the same outcome—luck. These tools facilitate critiques of the product, allowing for constructive changes to improve the product design.

Note

1. French, Simon, "Cynefin: Repeatability, Science and Values," *Newsletter of the European Working Group "Multicriteria Aid for Decisions,"* Series 3(No. 17), 1, Spring 2008. Available at: http://www.inescc.pt/~ewgmcda/OpFrench.pdf, (accessed June 19, 2010).

Chapter 9

Documenting the Testing

Documenting the testing aids the test engineering group in a number of ways (see Figure 9.1):

- Plans provide structure.
- Reports record results.
- Utilization and efficiency measurements provide data for quotations.

The first is the preparation work required for testing. The time to develop and design test fixtures is as early as possible, not after the product is delivered when it will push out the delivery date. Identification of areas requiring verification makes it possible for the team to determine if new equipment is needed and to identify performance requirements for that equipment.

The second benefit of documentation is that it choreographs the test group's response to the product test needs. Considering that development work seldom is contained and causes the testing portion of the project to be compressed, planning means the testing is down to the execution and responding to unexpected issues. Additionally, it is possible to prioritize the high-risk portions of the test—those cases in which a failure will be more than a quality blemish.

Next, documenting the testing early makes it possible to assess the coverage of the testing in advance of the actual testing. Use this documentation and time to critique what is believed to be the high-risk areas in terms of probability and consequence of any failure, allowing the adjusting of the testing scope to address these critical areas.

Documenting the test execution with a test report provides *traceability* of the actions conducted on the product. In the event that we have a problem not found in testing, but found at the customer's location, a critique of the testing activities will provide information on future testing needs. For example, consider an existing product undergoing a modification. The modification does not apply to all instances of the product but is based on customer demands. This specific product is

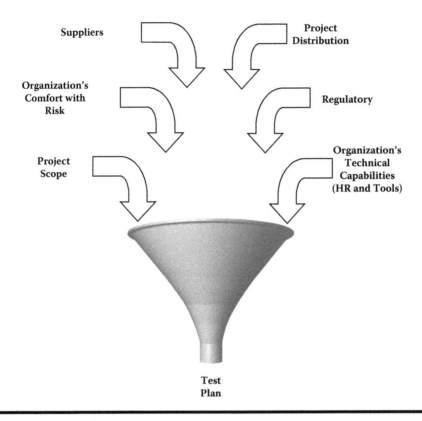

Figure 9.1 Test planning.

configurable at the end customer manufacturing facility, where a parameter is set in the component to enable the feature when selected by the customer. The test group verifies that the product modification performs as expected when placed in a vehicle configured with the system. The product goes to production and then performance issues showed up in the vehicles that did not have the system. The review of the testing documentation indicated that the variation of the product with the parameter disabled was not tested, and the test department adjusted for this in future testing.

9.1 Components of a Test Plan

9.1.1 Title

9.1.2 Purpose

The purpose of the test plan is one of the most important items in the documentation. We wish to assure ourselves that we are not "going through the motions" of testing— we would rather conduct testing that elicits useful information about the product.

9.1.3 Tester and Supervisor

We recommend identifying the test engineer/technician and the test supervisor for tracking purposes as well as providing the human context of the test execution. Some laboratories require sign-off from these individuals as an indication that they have seen and approved the final version of the test report.

9.1.4 Contact Information

We use the contact information to provide the telephone numbers and email addresses of the following individuals:

- Test engineer/technician
- Test supervisor
- Customer contact

This information is useful when questions arise regarding the prosecution of the test suite and the interpretation of the results.

9.1.5 Contents

The table of contents may seem mundane, but it can make life simpler for our customers, particularly if we are dealing with a very large test plan. We also verify the structure of our document with the table of contents.

9.1.6 Mechanical Tests

We specify mechanical tests with some simple parameters:

- Rise time
- Dwell time
- Fall time
- Number of cycles
- Maximum value
- Minimum value

These are the minimum values that should be specified. We would also expect to see in a test plan some support information describing the correct and incorrect behaviors of the unit under test.

Typical mechanical tests include, but are not limited to, the following:

- Vibration
- Mechanical shock
- Drop test
- Pendulum impact test

- Pull tests
- Push tests

9.1.7 Environmental Tests

We also specify environmental tests with some simple parameters:

- Rise time
- Dwell time
- Fall time
- Number of cycles
- Maximum value
- Minimum value
- Relative humidity
- Temperature

These are the minimum values that should be specified. We would also expect to see in a test plan some support information describing the correct and incorrect behaviors of the unit under test.

Typical environmental tests include, but are not limited to, the following:

- High humidity
- Immersion
- Chemicals
- Spray
- Steam
- Air pressure
- Frost
- Dust
- Pebbles
- UV exposure

9.1.8 Electronic Tests

We specify electronic and electrical tests with some simple parameters:

- Rise time
- Dwell time
- Fall time
- Number of cycles
- Maximum value
- Minimum value
- Relative humidity (a specific interval is allowed by certain standards)
- Temperature (a specific interval is allowed by certain standards)

These are the minimum values that should be specified. We would also expect to see in a test plan some support information describing the correct and incorrect behaviors of the unit under test.

Typical electrical/electronics tests include, but are not limited to, the following:

- Conducted emissions
- Radiated emissions
- Conducted immunity
- Radiated immunity
- Electrostatic discharge
- Chattering relays
- Positive/negative inductive transients
- Load dump
- Open circuit
- Short circuit
- Mutual coupling

9.1.9 Software Tests

Software tests run the gamut; however, there are some large categories of software testing:

- Black-box testing means we only know the inputs, outputs, and behavior of the units under test.
- White-box testing means we know how the "inside" of the box works.
- Unit, component, system integration, and system testing.

9.1.10 Measurement Uncertainty

Measurement uncertainty is a value much beloved by die-hard metrologists. It provides accuracy/precision information about the instruments we are using to measure the results of our tests. Very coarse uncertainty is only usable with very coarse tests, whereas a minute level of uncertainty is usable with all tests. Regardless, our equipment information should include the measurement uncertainty budget for each measuring device.

9.1.11 Equipment Calibration Dates

For this item, we either list the most recent calibration date or the next calibration date. The latter version is often preferred by metrology gurus because it is slightly easier to get a fix on whether or not we are overdue. Each of these required data force us to ensure that our measuring equipment is well tended to.

9.1.12 Equipment Identification

Each piece of our test equipment and measuring devices should be uniquely identified—this goal is often accomplished with inventory tags. The particular method is irrelevant as long as the hardware has some identification. If we have a suspect piece of equipment, it should have a name and location. Customer-owned equipment should also be identified because it will actually belong on their inventory and not ours.

9.1.12.1 Brand/Type

We want to specify everything we can about the brand, model, type, version, and age of our equipment. All of this bookkeeping shows our customer and the rest of the world that we know what we test, what we tested with, when, and so on. If our organization ever has to go to court over a product issue, the more detail we can present, the more credible we will be.

9.1.12.2 Inventory Number and Instructions

We mentioned inventory identification earlier, but we also want to keep our operating instruction/manuals as controlled documents whenever possible. That means that our document configuration management department—if we have one—knows the location of each document and can retrieve them when needed to replace them with updated documents.

9.1.12.3 Limits

We should list the manufacturer's operation limits for each device used for a test. Clearly, we want the test range to lie within those limits. Furthermore, we need to ensure that our equipment calibration covers the expected range of use for the equipment. We do not always expect the calibrated range to be identical with the equipment range; however, it should *never* exceed the equipment range (exceeding the bounds makes no sense whatsoever).

9.1.12.4 Failure Criteria

Somewhere in our test document we should explicitly call out the observable failure criteria for the unit under test. Mere familiarity with the product is not really satisfactory—we need objective criteria. Even better is if we can implement our criteria using automated testing, allowing us to make the capture of the anomalies even more objective. By taking this step, we endeavor to eliminate anything that resembles an opinion.

9.1.12.5 Physical

We can provide physical measurements (e.g., length, width, height) for each sample part as well as for the environmental chambers if we are doing that type of testing. If we are testing multiple units in a given chamber, we need to ensure that we are not having shadowing issues, where one unit blocks the effect on another unit. This situation can occur especially in tests where we use moisture or chemical sprays. Each unit under test should receive the same level of attack as any other unit.

9.1.12.6 Performance

We should verify the performance of our equipment regularly as a quick check on the capability of the equipment. Clearly, calibration would be a more exacting approach, but it is also more expensive and time consuming. We can quick-check environmental chambers by seeing how long it takes to heat a predetermined amount of water to a specific temperature.

9.2 Components of a Test Report

9.2.1 Same as Test Plan + Description of Outcome

If we design our test plan correctly, we can add information to a result column and immediately have the bulk of a test report completed. Note that we said the "bulk" of a test report. We find it prudent to add more information, which we describe in the next few subsections.

9.2.1.1 Photos Welcome

In any case where a failure is visible, either to the eye or through a microscope, we recommend taking photographs of the observed failure mode. These pictures will often put a damper on inane responses from clients or customers. Furthermore, the photographs add another layer of objective description—the picture is usually more concise than the accompanying verbiage.

9.2.1.2 X-ray If Applicable

In any case where a failure is not visible, either to the eye or through a microscope, we recommend taking x-ray photographs of the observed failure mode. These pictures will also often put a damper on inane responses from clients or customers. Again, the x-ray pictures add another layer of objective description—although they may require more explanation than simple photographs of the failure mode. One typical example of an unseen failure mode occurs when an electronic part (like a microcontroller) fails from electrostatic discharge. The actual damage cannot usually be seen by the eye or even by x-rays. The parts will usually be sent to the supplier/manufacturer for delamination and then visual examination followed by a photomicrograph.

9.2.1.3 Failure Description If Any

We usually provide at least a short textual description of the failure, particularly if failures are logged in a data base. If we have a well-developed taxonomy for failures, we can use the description from the taxonomy as part of the failure description. Both IEEE and Boris Beizer provide taxonomies for describing software failures. We see no reason why we can't develop a taxonomy for hardware failures as well.

9.2.1.4 Observations If Any (If Interesting Non-Failure or Non-Relevant Failure Event Occurred)

Even if the observed item is not a relevant failure, we generally recommend that these events be recorded as observations. If we are using a database, we might set a field for follow-up on the off chance that the observation describes an issue relevant to another part of the project. We have also seen laboratories log observations as internal test incidents for the same reason.

9.3 The DVP&R Format

DVP&R stands for design verification plan and report. Each automotive firm has its own format for this document, although they all bear a family resemblance. The format provides a concise summary of testing activities as well as combining the idea of a plan and a report into one document. We can use the DVP&R for both hardware and software testing. The example in Figure 9.2 represents a suite of hardware tests conducted against a sample of an automotive product. And the example in Figure 9.3 represents a suite of software tests conducted against a sample of an automotive product with the appropriate product software embedded in the microcontroller's flash memory.

Because the DVP&R format is so concise, we will normally provide more detailed documents in support of it. However, the DVP&R provides a quick, high-level overview of what transpired during the execution of the test suite and is thus suitable for management review also.

9.4 Failures

9.4.1 Relevant Failures (MIL-STD-2074)

U.S. Department of Defense Military Standard 2074 describes some rules for determining whether failures are relevant or irrelevant. Because the decision points are rational, we recommend this document to any test group as a means of defining these situations.

	Product Validation Plan and Report (DVPR - PVPR Combined)						
Product Name:	Customer Name:						Electrical Engineer:
This product	*This customer*						*This EE*
Last PVPR Update:	Stoneridge P/N:			Project ID:			Mechanical Engineer:
This date	*TBD*			*TBD*			*This ME*
		# of Samples			Test Timing		
Test Name	Specification	Planned	Actual	Test Results	Sched	Actual	Comments/TIRs
Product Version: A, Software Version: 1.0, This Customer Part #: 123456789							
Test Requestor: This program manager Reason for test cycle: DVT							
Short Circuit to Ground	522-032XXX-0001	3			10/04/05-10/07/05	TBD	
Short Circuit to Battery	522-032XXX-0001	3			10/04/05-10/07/05	TBD	
Short Circuit to Zero Volt Reference	522-032XXX-0001	3			10/04/05-10/07/05	TBD	
Open Circuit	522-032XXX-0001	3			10/04/05-10/07/05	TBD	
Cranking Motor Voltage	522-032XXX-0001	3	2 (11-12)	Pass	10/04/05-10/07/05	10/05/05	
Failed Alternator Voltage	522-032XXX-0001	3	3 (10-12)	Pass	10/03/05-10/04/05	10/03/05	

Figure 9.2 Hardware design verification plan and report (DVP&R).

Full Field Alternator Voltage	522-032XXX-0001	3	3(10-12)	Pass	10/03/05-10/04/05	10/04/05	
Batteryless Operation	522-032XXX-0001	3			10/04/05-10/07/05	TBD	
Jump Start Voltage	522-032XXX-0001	3	2(11-12)	Anomaly	10/03/05-10/04/05	10/04/05	TIR # 213
Reverse Battery	522-032XXX-0001	3	2(11-12)	Pass	10/03/05-10/04/05	10/04/05	
Mutual Coupling	522-032XXX-0001	3			10/14/05-10/20/05	TBD	
Negative Inductive Spikes	522-032XXX-0001	3			10/14/05-10/20/05	TBD	
Switching Spikes	522-032XXX-0001	3			10/14/05-10/20/05	TBD	
Load Dump Transient	522-032XXX-0001	3			10/14/05-10/20/05	TBD	
Chattering Relay	522-032XXX-0001	3			10/14/05-10/20/05	TBD	
Electrostatic Discharge	522-032XXX-0001	3	3(14-16)	Pass	10/10/05-10/20/05	10/11/05-10/13/05	
Conducted Emissions	522-032XXX-0001	3			10/10/05-10/11/05	TBD	
Radiated Emissions	522-032XXX-0001	3	3(17-19)	In progress	10/12/05-10/13/05	10/10/05	TIR # 217
Radiated Immunity	522-032XXX-0001	3			10/13/05-10/14/05	TBD	
Current Loading Test	522-032XXX-0001	3		n/a	TBD	TBD	
Insulation Test	522-032XXX-0001	3		n/a	TBD	TBD	
Leakage Current Test	522-032XXX-0001	3		n/a	TBD	TBD	
Voltage Drop Test	522-032XXX-0001	3		n/a	TBD	TBD	

Figure 9.2 (Continued).

Test	Part Number			Result			Notes
Chemical Compatibility	522-032XXX-0001	3			TBD	TBD	To Be Outsourced or Prior Product Results
Corrosives	522-032XXX-0001	3			TBD	TBD	To Be Outsourced or Prior Product Results
Ultraviolet Resistance	522-032XXX-0001	3			TBD	TBD	To Be Outsourced or Prior Product Results
Flammability	522-032XXX-0001	3			TBD	TBD	To Be Outsourced or Prior Product Results
Temperature Ranges & Voltages	522-032XXX-0001	3	3 (4-6)	Pass	09/23/05	09/23-09/24/05	
Thermal Shock	522-032XXX-0001	3	3 (4-6)	Pass	09/26-09/27/05	09/26-09/27/05	
Humidity	522-032XXX-0001	1	1 (# 4)	Pass	09/29-10/01/05	09/29-10/01/05	
Humidity	522-032XXX-0001	1	1 (#5)	Pass	10/01-10/03/05	10/01-10/03/05	
Humidity	522-032XXX-0001	1	1 (#6)	Pass	10/03-10/05/05	10/03-10/05/05	
Storage Temperature	522-032XXX-0001	3	3 (1-3)	Pass	09/22-09/30/05	09/22-	
Temperature Cycle	522-032XXX-0001	1	1 (#1)	Pass	10/05-10/14/05	10/05-10/14/05	
Temperature Cycle	522-032XXX-0001	1	1 (#2)	In progress	10/14-10/23/05	10/14-	
Temperature Cycle	522-032XXX-0001	1	1 (#3)		10/24-10/26/05	TBD	
Frost Test	522-032XXX-0001	3	3 (1-3)		10/26-11/02/05	TBD	

Figure 9.2 (Continued).

Test	Part Number	Qty	Units	Result	Date	Date	Notes
Splash	522-032XXX-0001	3	3 (4-6)		10/10-10/11/05	TBD	
Steam Cleaning	522-032XXX-0001	3	3 (7-9)	Anomaly	09/30/05	10/04/05	See TIR # 214.
Pressure Washing	522-032XXX-0001	3	?		TBD	TBD	**Need to confirm if applicable**
Dust Test	522-032XXX-0001	3	3 (7-9)	Pass	09/29/05	09/29/05	
Salt Spray	522-032XXX-0001	3	3 (7-9)	Anomaly	10/05-10/10/05	10/05-10/10/05	NOTE: Tested only unit #7 & 9 due to have only 2 set of connectors available and unit #8 failed during steam test. See TIR # 219 & 220.
Water Proofing	522-032XXX-0001	3	3 (7-9)		10/03/05	TBD	
Spill Test	522-032XXX-0001	3	3 (4-6)		10/08/05	TBD	
Random Vibration	522-032XXX-0001	3	3 (7-9)	Pass	09/23-09/28/05	09/23-09/28/05	
Swept Sine Vibration	522-032XXX-0001	3	?		TBD	TBD	**Need to confirm if applicable**
Handling Drop	522-032XXX-0001	3	3 (4-6)		10/7/2005	TBD	
Transit Drop	522-032XXX-0001	3			TBD	TBD	**Need to packaging assembly.**
Reliability	522-032XXX-0001	TBD			TBD	TBD	

Figure 9.2 (Continued).

Design Verification Plan and Report

SOFTWARE ENGINEERING SIGNOFF:			Created:		Jaime Yañez		Revised:
System **This product**	Assembly:		Program:				9/28/2005
Specification No: *This specification*			Freightliner "FCCC"			Design Engineer *This software engineer*	
Subsystem All	Software Version 0.994		Latest Design Level			Concurred *This test engineer*	

Test Name/Source	Acceptance Criteria	Test Results	Design Level Tested	No. test cases		Software Test Description	Test Timing		Remarks
				Required	Tested		Sched	Actual	
STR									
Functional Test									
Unless otherwise stated, all tests are Our Test Team.									
N/I : Not Implemented									
SOFTWARE TESTS:									
1.0 CAN Handler	Fail	44	*SVT	125	125	J1939& 1708 protocols	23-Sep-05	23-Sep-05	1.4-1.7,1.15-1.18,1.43-1.45, 1.47-1.68,1.115-1.125

Figure 9.3   Software design verification plan and report (DVP&R).

2.0 J1939	Fail	26	*SVT	78	72	J1939& 1708 protocols	23-Sep-05	23-Sep-05	2.6,2.9,2.11,2.12,2.21-2.27, 2.43,2.45,2.49,2.50,2.52, 2.53,2.59-2.64,2.71,2.72, 2.75,2.76,
3.0 J1708 handler	Fail	5	*SVT	22	16	J1939& 1708 protocols	23-Sep-05	23-Sep-05	3.2-3.4,3.6,3.8,3.10
4.0 J1587	Fail	18	*SVT	38	34	J1939& 1708 protocols	23-Sep-05	23-Sep-05	4.3,4.4,4.6-4.8,4.14,4.15, 4.17-4.21,4.27-4.32
1.0 Odometer	Faults	2	*SVT	82	74	ODOMETER	24-Sep-05	24-Sep-05	1.33,1.74
1.0 SPEEDOMETER	Faults	25	*SVT	94	94	Gauges	25-Sep-05	25-Sep-05	1.69-1.93
2.0 Tachometer	Fail	1	*SVT	39	39	Gauges	25-Sep-05	25-Sep-05	2.23
3.0 Engine Coolant	Faults	6	*SVT	24	24	Gauges	25-Sep-05	25-Sep-05	3.23,3.8,3.10,3.14,3.20,3.23
4.0 Fuel Level	Pass	Pass	*SVT	55	38	Gauges	25-Sep-05	25-Sep-05	
5.0 Primary Air System Pressure	Faults	15	*SVT	24	24	Gauges	25-Sep-05	25-Sep-05	5.1-5.13,5.22,5.24
6.0 Secondary Air System Pressure	Faults	15	*SVT	24	24	Gauges	25-Sep-05	25-Sep-05	6.1-6.14,6.22,6.24

Figure 9.3 (Continued).

7.0 Engine Oil Pressure	Pass	Pass	*SVT	28	28	Gauges	25-Sep-05	25-Sep-05	
8.0 Transmission Oil Temp	Pass	Pass	*SVT	10	10	Gauges	25-Sep-05	25-Sep-05	
9.0 Turbo boost pressure	Pass	Pass	*SVT	14	14	Gauges	25-Sep-05	25-Sep-05	
10.0 Battery voltage	Pass	Pass	*SVT	26	26	Gauges	25-Sep-05	25-Sep-05	

Figure 9.3 (Continued).

9.4.1.1 Design and Workmanship Failures

Failures due to design deficiencies or poor workmanship of either the system or component parts are decidedly relevant, particularly in the automotive testing arena when we are performing production validation testing.

9.4.1.2 Component Part Failures

If several component parts of the same type or designation fail during any test, each one is a separate relevant failure unless it can be shown that one failure caused one or more of the others (cascading failures).

9.4.1.3 Wear-Out Parts

Certain parts of known limited life, such as batteries, may have a well-known life before the start of testing. In this case, we do not want to call the result a test failure when we already knew the battery would cease to function and the battery is *not* the unit under test. If the battery, for example, fails before the *expected* failure time, we have at least an observation, if not a failure mode. On the other hand, the failure of one of these parts that happens after the known operational period is essentially irrelevant, but any dependent failures observed may be relevant.

9.4.1.4 Multiple Failures

When concurrent unit failures happen, each failed unit that would independently prevent satisfactory product performance counts as a relevant failure except when

1. The failure of one part was entirely responsible for the failure of any other parts—in which case, each associated dependent part failure is *not* considered to be a relevant failure (does not add to failure count).
2. At least one unit relevant failure must be counted when we claim a dependent failure; this makes sense because we observed an issue with the product.

9.4.1.5 Intermittent Failures

The first occurrence of an intermittent (aperiodic) failure on any unit will be considered a relevant failure, although subsequent occurrences of the same intermittency on that same unit will be considered nonrelevant. In other words, once we identify the failure mode, we have identified the failure mode. However, we do not want to be naïve with intermittency—we must also identify the aperiodicity of the failure mode because that may be a key factor in the analysis and correction of the defect.

9.4.2 Irrelevant Failures (MIL-STD-2074)

9.4.2.1 Failures Directly Attributable to Improper Installation

Failures attributable to improper installation only become relevant failures when we are using our test sequence to *detect* improper installation or assembly. The key to this assessment lies with the purpose of the test: Are we testing the design? Or are we testing manufacture and assembly? The answers to these questions determine the relevancy of the failure.

9.4.2.2 Failures of Test Instrumentation or Monitoring Equipment

We are testing the units under test, not the test equipment per se. Hence, the failure of test instrumentation or monitoring equipment should not count against the unit under test. For example, we have seen an antenna destroy itself while doing radiated immunity at a field intensity of 200 V/m, which is incredibly high. The failure of the antenna had nothing to do with the unit under test.

9.4.2.3 Failures Resulting from Test Operator Error in Setting Up or Testing the Equipment

This irrelevancy is similar to the equipment issue. We can record these incidents somewhere because it is possible that we have a systemic procedural issue with our testing group.

9.4.2.4 Dependent Failures, Unless Caused by Degradation of Items of Known Limited Life

Dependent failures are exactly that—dependent failures. We are usually interested in the primary failure. However, if the sequence of dependent failures is catastrophic, we may be seeing a lack of design robustness in the product.

9.4.2.5 Count at Least One Relevant Failure When Claiming a Dependent Failure

If we have a dependent failure, then we must have an independent failure. Therefore, we count at least one failure based on the independent failure. Again, the system makes sense.

9.4.2.6 Failures Attributable to an Error in the Test Procedures

An error in the test procedure is similar to errors from the operator or from the equipment. The unit under test is not causing the problem—something external to the unit under test becomes an issue. This problem explains why every test is a test of the test as well as a test of the sample unit.

9.4.2.7 The Second and Subsequent Observations of the Same Intermittent Failure on the Same Unit

While we want to note the intermittency because it may be a clue to the failure mode, subsequent observations of the same failure do not constitute a new failure and are thus irrelevant. In essence, we are interested in logging *unique* failures.

9.4.2.8 Failures Occurring during Any Pre-Test Burn-In, Trouble-Shooting, Repair Verification, or Setup Time

We wish to record these failures as an observation for future reference if needed. However, none of these activities is the actual test case. We record the observation to see if systemic issues with the product arise.

9.4.2.9 Failures Clearly Assignable to an Overstress Condition Outside the Design Requirements and Outside the Scope of That Particular Test

This failure may be irrelevant. It depends on the purpose of the test: If we are doing product characterization, we want to elicit the failure limits and the destruction limits. During highly accelerated life testing (HALT), we deliberately introduce overstress conditions in order to encourage the product to fail at its weakest point.

Note

1. French, Simon, "Cynefin: Repeatability, Science and Values," *Newsletter of the European Working Group "Multicriteria Aid for Decisions."* Series 3(No. 17), 1, Spring 2008. Available at: http://www.inescc.pt/~ewgmcda/OpFrench.pdf (accessed June 19, 2010).

Chapter 10

Test Administration

Test administration covers all the actions that are employed to deliver the testing results. This includes identifying the test needs, preparing for those tests, executing the tests and fault recording, and tracking any necessary corrective actions.

- Develop test equipment and methods.
- Optimize test throughput.
- Report test results diligently.
- Ensure that necessary test systems (fault reporting and bug tracking, etc.) are in place.
- Develop the human resources within the group.

Figure 10.1 depicts the results of a fault reporting process over time. Note that the figure approximately follows a Rayleigh distribution, a vagueness typical of real data.

10.1 Test Management

Like any other kind of management, test management is comprised of

- Planning
- Organizing
- Staffing
- Delegating
- Supervising
- Measuring
- Reporting

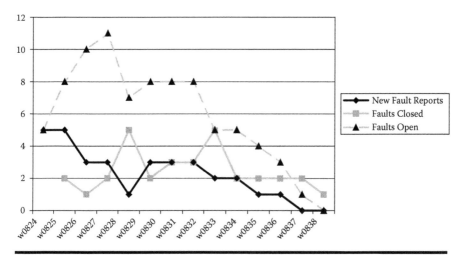

Figure 10.1 Fault reporting.

Consequently, our key metrics will be related to these seven items. The testing organization can plan for both the short term and the long term. A short-term plan might be good for a planning horizon of a month. A long-term plan, however, can extend out as far as makes sense. One of us manages a lab where the long-term plan extends out for a year. We publish the plan to project managers, commercial managers, engineering departments, and production departments. The long-term plan gives a broad-brush view of expected laboratory resource consumption and gives other parts of the organization a chance to plan their work during the less-occupied periods. The short-range plan, on the other hand, provides the laboratory with enough detail to adequately test within a more obvious future timebox.

If the testing organization operates under ISO/TS 16949 or ISO/IEC 17025, then they will have an organization designed to comply with these standards. If calibration of equipment is an organizational responsibility, then they should be working under ISO/IEC 17025 and ANSI Z540.3:2006 (at least in North America). Other than meeting requirements in the standards, the test group might be set up as follows:

- The typical hierarchical pyramid
- A self-directed work team
- Decentralized work circles

The organization will most likely be driven by corporate requirements as well as exigencies of the business in which the test group operates.

As always, staffing is a key component of successful test management. In one of our organizations, the test laboratory is primarily staffed by engineers rather than technicians. The engineers are capable of writing software (Labview), performing detailed analyses, and driving continuous improvement. In other situations, we might see a technician-centric test staff, particularly when a customer brings driver software to the test facility.

We expect the test group manager to delegate tasks to the persons most able to perform these tasks. If we are using a self-directed work team, we might see the team discuss allocation of tasks. In some cases, we will assign tasks that allow employees to experience growth in skills.

Supervising occurs whether we have a pyramidal organizational scheme or a working circle. In one case, the titular head of the organization supervises; in the other cases, consensus represents self-supervision. Sometimes, a quality standard or a strong customer requirement may force the type of supervision. For example, if a laboratory is pursuing the North American Automotive Electromagnetic Compliance Laboratory Recognition Program (or AEMCLRP) designation, it will be required to have a National Association of Radio and Telecommunications Engineers (NARTE)-certified electromagnetic compliance (EMC) engineer on site.

The testing organization has a responsibility to measure critical performance indices to determine if they are improving or degrading as a team. Typical measurements can be as follows:

- Test incident reports
- Closure time for test incident reports
- Testing resource loading
- External service income
- Equipment maintenance costs
- Budgeted expenses
- Unbudgeted expenses
- Calibration activities
- Items past due

We would expect these items to be reported at least monthly or cycle with the accounting department reporting frequency. Of course, as part of the testing activities, we will also measure failures, voltages, currents, temperature, and any other variable that is appropriate to the test.

We not only provide test reports, but we will also report on our metrics as an organization. The test group might consider some kind of dashboard or, perhaps, simply post their graphs, tables, and other metrics on a bulletin board to be updated at meaningful intervals.

If our test group is testing software, we would expect most of the same items to be significant. We might add metrics for defect containment, overall defect discovery rates, and defect escapes that reach the customer. We may choose to associate the testing activities and number of fault reports per test engineer as a way of focusing attention on developing those specific areas.

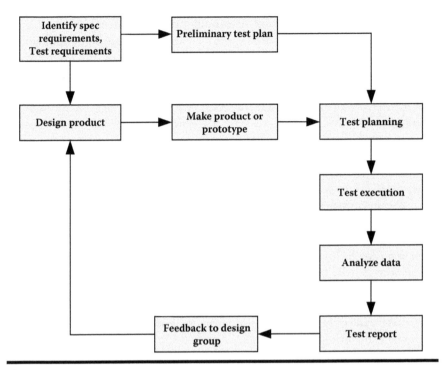

Figure 10.2   Test management process.

10.1.1 Test Process

A major goal is to use our test process to improve the capability of the organization. We can do this by considering the following items:

- Performance:
  - Priorities
  - Capabilities
  - Functions
- Efficiency:
  - Maintenance
  - Logistics
  - Operations
- Availability:
  - Support
  - Maintenance (down time)
  - Reliability (utilization)

Note how our list roughly corresponds with the common manufacturing concept of overall equipment effectiveness, which is the product of performance, efficiency, and availability with some quality concepts poured into the stew. The testing process

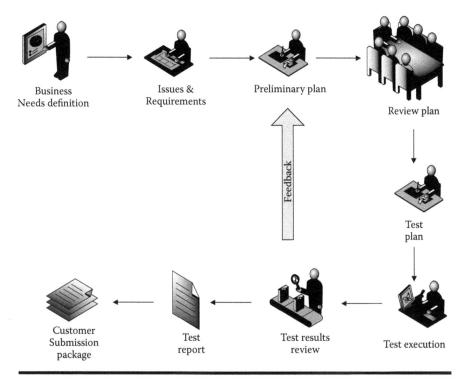

Figure 10.3   Test and correction process.

should be defined long before the actual testing starts. Understanding the scope of what is to be tested, prioritizing the risky areas, and planning how to achieve a level of confidence that is acceptable takes time.

A typical automotive process will include design verification and product validation as major components. The U.S. Department of Defense considers developmental testing followed by operational testing, which we think is a completely logical approach to the testing process. Figure 10.3 shows an example of a test process (certainly not the only possibility). The feedback part of the process allows for continuous improvement in laboratory practices.

10.1.2 Test Marketing

If a test group decides to pursue business internally or externally, it must market itself. Here are some items we have found to be relevant to the marketing/sales process:

■ Prospecting
■ Qualifying
■ Capabilities
■ Establishing rapport and trust

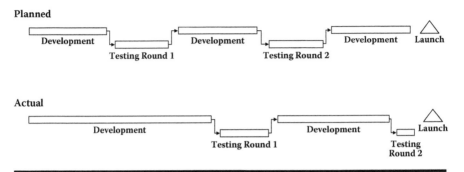

Figure 10.4 Test planned schedule versus actual schedule.

- Identifying needs or problems
- Presenting solutions
- Answering objections
- Closing the sale
- Getting resales and referrals

10.2 Test Scheduling

Anybody who has been in the testing environment for long knows the challenges of testing and schedules. No offense to those development people, but they can take much more time to develop the product than the test group has to verify and validate the product. The test team gets the remainder of the schedule. Seldom does the organization entertain any delay of the product launch. Sometimes the enterprise will have compelling reasons, such as legal demands, that necessitate maintaining the schedule. This compression of test times generates opportunities to improve the coverage and efficiency of the tests for the product. Figure 10.4 shows the "scrunching" effect that occurs as schedule slips happen.

10.2.1 Schedule the Tests as Early as Possible

Outside laboratory sources, such as external laboratories or equipment rental firms, can have long lead times. Even for internal testing, access to expensive equipment that we want to use may be in nearly continuous use. This equipment can be quite expensive to purchase or rent or may not even be available (we have seen this happen with a Schaffner transient tester). The business case to purchase is based on a certain use rate, and that usually translates to nearly continuous use of equipment before we are allowed to purchase new equipment. Hence, we require early enough scheduling to be able to inform our clients about the schedule and to negotiate new schedules. The problem is exacerbated when the outside lab is the only place to perform the work; for example, a complete 18-wheel vehicle (class 8 truck tractor) with EMC requires a chamber large enough to hold the vehicle.

10.2.2 Effort Is Not Labor Intensive

Commonly, mechanical and environmental testing requires setup by an engineer or a technician who can then walk away from the device. We record the results using closed-circuit cameras and store what we see on DVD or hard drive. Using this approach, we can provide full-day service without requiring the presence of a human. If we have any questions regarding the results, we can burn a DVD or use some other storage mechanism and send the time-stamped video to the customer to review and comment.

To plan for this environment, we are more interested in machine availability than we are in human resource availability. In short, we are machine constrained. When using a tool such as Microsoft® Project, the machine is the resource—not the person.

10.2.3 Effort Is Labor Intensive

EMC testing is usually labor intensive. Examples of these tests are as follows:

- Radiated susceptibility
- Conducted susceptibility
- Radiated emissions
- Conducted emissions
- Bulk current injection
- Electrostatic discharge
- Electrical transients:
  - Positive and negative inductive spikes
  - Load dump
  - Mutual coupling
  - Chattering relays

In this situation, we are usually more concerned about the availability of the human resource than we are about machine availability. Electrostatic discharge tests require substantial attention if we are attacking a product with, for example, 90 input/output pins that must be tested in a special sequence and "zapped" every 5 seconds!

10.2.4 Do Not Wait Until Product Release to Schedule Testing

We issue short-range and long-range plans to provide enough information for other parts of the organization to intelligently use the test team as a resource. Under no circumstances would we wait until product release to schedule testing, even if we are only discussing prototype product releases. Any time we have to rush our work, we increase the probability of poor craftsmanship.

10.3 Test Human Resources

Testing, like any other aspect of product development, hinges on the abilities of the people in place to perform the work.

10.3.1 Technicians versus Engineers

10.3.1.1 Engineers Can Write Test Code

In our experience, most third-party labs staff their teams primarily with technicians. Third-party labs will usually require the presence of their customers if any special stimuli must be performed on the test samples. We have generally supplied a design engineer who completely understands the product. If testing at the third-party site consumes a lot of this engineer's time, we may not be making the best use of our human resource.

10.3.1.2 The Laboratory Uses the LabVIEW Product

We use the National Instruments LabVIEW product exclusively. We use this approach not so much because of the graphical programming that is part of the LabVIEW technique, but more because of the excellent hardware support also provided by National Instruments. Basically, we are able to create a fairly complete test bench for electronics testing on a single computer. In Figure 10.5, the item marked "Sequencer" is a National Instruments chassis replete with a variety of circuit interfaces required for the device under test.

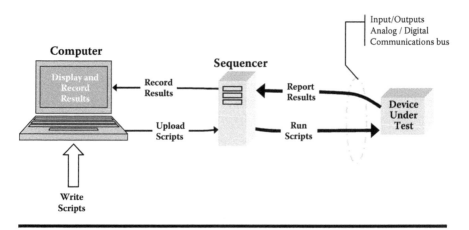

Figure 10.5 Test bench.

This fixture is essentially a state machine, allowing the test staff to write a sequence of commands. The commands themselves are action words that reflect the actions that the team will take while using the fixture. These commands, or *script*, can be performed repeatedly and automatically—increasing test coverage. When requirements change, the test script or commands will change. This is usually an efficient way to address regression testing. Figure 10.6 is an example of the output from such a test fixture—displaying the tests conducted and the results.

10.3.1.3 Engineers Can Provide a Level of Analysis That a Technician Cannot

Conducting testing by employing engineers instead of technicians facilitates complex testing because the engineers are better armed with the mathematics, physics, programming, and other technical skills. We are able to optimize our test planning due to the fact that engineers conduct the testing. The higher skill level of the engineer allows them to read what the test results are telling them and move accordingly to squeeze a large amount of information from the product.

10.4 Test Device Resources

10.4.1 Whenever Possible, Provide Device Backup

Backup devices provide a contingency for equipment breakdowns. If we are pushing lab capacity, we want to ensure that we do not have any stoppages. Of course, some may say that we are not at capacity if we have backup equipment; however, in some cases we may have equipment that is not regularly used because of calibration issues, archaic interfaces, or difficulty setting up.

10.4.2 Keep the Devices Calibrated

It is not possible to make an informed decision about the test results if these results have a large variance, which will amount to comparing one device of an unknown measurement uncertainty with another device of an unknown measurement uncertainty. Testing a product with a piece of test equipment that has an unknown measurement uncertainty does not provide much, if any, useful information regarding the product and certainly not much reliability in the test results.

10.4.3 A Slower Device May Work Well as a Backup

Sometimes, in order to accomplish our schedule, we may need to use older tools or slower tools to get the job done. Examples of these kinds of tools are the following:

■ Older personal computers
■ Analog scopes

Step	Action	Type	Name	Value	Operator	Update rate [ms]	Status
1	Set	Battery	BAT+	13.00 V;1.00 A			Pass
2	Wait			2			Pass
* End of unit_startup.seq *							
3	Wait			2			Pass
					C:\simulator\apci\NEW MAP2\hybrid.map;		
4	Set	J1587 Parameters	EA;ZU;ZV	0;1;0	C:\simulator\apci\NEW MAP2\dst1.hex;		
					C:\simulator\apci\NEW MAP2\dst2.hex		Pass
5	Wait			2			Pass
6	Set	Battery	BAT+	0.00 V;1.00 A			Pass
7	Wait			2			Pass

Figure 10.6   Test report.

			BAT+	13.00 V;1.00 A			Pass
8	Set	Battery					Pass
9	Wait			2			Pass
	* End of unit.startup.seq *						
10	Wait			2			Pass
11	Get	Digital	Engine_Remote_Start_Enable	1			Pass
12	Wait			2			Pass
13	Set	J1939	EngineTestModeSW	1	CAN0	100	Pass
14	Set	J1939	EngineTorqueMode	0	CAN0	100	Pass
15	Wait			2			Pass

Figure 10.6   (Continued).

16	Get	Digital	Engine_Remote_Start_Enable	0				Pass
17	Check				0.00±0.00	Equal		Fail
18	Set	J1939	EngineTestModeSW	0		CAN0	100	Pass
19	Wait			2				Pass
20	Get	Digital	Engine_Remote_Start_Enable	1				Pass
21	Check				1.00±0.00	Equal		Pass
22	Clear	J1939	CCVS_X_V					Pass
23	Clear	J1939	EEC1_X_E					Pass

SUM: FAIL

Figure 10.6 (Continued).

- Alligator-clip adapters (built on the spur of the moment and checked out!)
- Large power supplies stepped down (transformed) to behave as multiple smaller power supplies

The message here is that we do not want to delay a test simply because we do not have the latest equipment. We see no reason that we cannot improvise some of our equipment before testing—as long as we understand the ramifications to measurement and test results.

10.4.4 Consider Alternative Methods for Testing

10.4.4.1 Bulk Current Injection Substitutes for Antennas in Radiated Susceptibility Testing

When testing up to 400 MHz, we can sometimes substitute bulk current injection (BCI) for antenna testing. To substantiate the substitution, we should already have performed correlation tests in our chambers with known units under test.

10.4.4.2 Another Lab

We can always send our sample units to another laboratory or proving ground. In general, we recommend sending the parts with one of our own development engineers and the appropriate fixtures. If we work under a quality operating system such as ISO/TS 16949, we will need to ensure that the external laboratory is ISO/IEC 17025 compliant.

10.4.4.3 Simulation

Usually, simulation is only a realistic choice for testing when we are in the early phases of development. We can still use the simulators later in the development if we understand their limitations. A simulator can also be a quick way to test out an idea.

10.4.4.4 Similar Product

We may be able to learn most of what we need to know by testing a variant of our product or a similar product. Obviously, we would note the deviation on the test report.

10.5 Test Quality

10.5.1 Issues Discovered Post-Testing

Effective testing will mean fewer faults found in the product by your customer. We maintain metrics that indicate the number of post-launch problems detected in the product, especially for software versions, which are often more prone to field issues.

We have also seen what we call *post-launch jitters*, particularly with warranty issues. This period typically lasts from roughly 6 weeks to approximately 6 months after launch. The following list illustrates some topics that can cause post-launch jitters:

- The switch from MRO buying (spot, non-MRP buying) to MRP-based buying of materials
- The use of alternate materials from a deviation
- Inadequate knowledge of how to build the product (supplier)
- Inadequate knowledge of how to install the product (customer)
- Inadequate knowledge of how to use the product (end user or operator)
- Rushed and late deliveries

We assume that most suppliers would like to eliminate the post-launch jitters in the interest of reducing warranty issues, returned merchandise, and dissatisfied customers. We recommend looking at each item in the list to ensure that we have plans in place to manage these issues.

10.5.2 Deliberately Adding Faults to the Unit under Test

Fault seeding is deliberately and randomly placing defects within the product before the testing and verification activities. The number of faults is not divulged to the test and verification group. The ratio of faults found to faults seeded provides an indication of the number of faults that may remain in the software. In theory, finding all the seeded faults is deemed a good indication that all faults may have been found. If our team does not discover all the seeded faults, then we would suspect our test suite is not functioning particularly well. In essence, we are providing one system for providing an estimate of the quality of the test suite.

$$N = \frac{F_s \times F_f}{N_s}$$

where:
N_s = Faults seeded
F_s = Number of faults seeded that are found
F_f = Total number of faults found
N = Estimated total number of faults

10.5.3 Testing Organization Should Always Be Prepared to Present Their Quality Data

The test team should be keeping metrics regarding the testing process, whether hardware or software. Typical metrics can include the following:

- Count of test incident reports
- Count of audit nonconformances

- Count of open corrective actions
- Total count of corrective actions
- Count of eight discipline documents
- Count of issues found after product part approval process (PPAP)
- Count of issues found after official launch
- Count of repeat problems (defect containment)

10.6 Test Costing

10.6.1 Calculate the Cost of Testing—The Amount of Time It Takes

The money spent on each defect can be estimated by the equation below:

$$CPF = \frac{C}{F}$$

where:

C = Money spent testing
F = Total number of faults found
CPF = Estimated money spent per fault found

10.6.2 Laboratory Costing

Figure 10.7 shows a fictitious example of a lab calculation using the Kaplan and Anderson approach. Their simplified method of activity-based costing provides, at a minimum, a sanity check for any other calculations. If we are confident in our numbers, we can use the results as they are in order to build our quotations.

10.6.3 Test Cost Quotation

If we know the cost per minute and we know the number of minutes the test takes, then we know the cost of the test, at least in a general sense. As with all accounting activities, we run into difficulties when trying to estimate real labor. For example, execution of an electrostatic discharge (ESD) test is extremely labor intensive because, in its usual form, the test requires stimulation on every pin of the connector every 5 seconds. If we have many pins, the work becomes tedious. On the other hand, if we are putting parts in environmental chambers and checking them at an hourly frequency or with cameras, the labor to execute the test is relatively minimal.

To avoid gouging our customers, we walk a fine line when quoting new projects. We can use our activity-based method as a check on anything we get from the accounting department and some experience and judgment.

Vacation accrual:

1year	10 days
2years	12 days
3years	12 days
4years	14 days
5years	14 days
6years	16 days
7years	17 days
Every 4 years thereafter	1additional day

Sick/personal days = N/A
Day without pay or if the Seguro Social can give them a leave voucher

Holidays	10 holidays observed	
	Vacation	Holidays
MR	12	10
GS	32	10
ER	18	10
DV	12	10
CM	12	10
JB	12	10
AH	15	10
VV	14	10
EL	14	10
AV	12	10
Total	153	100
Average	15.3	10

total work days - holidays	249
less average vacation days	−15.3
Real working days	233.7
Preliminary working hours	21033
Preliminary working minutes	1261980
Less meals	−141000
Actual working minutes	1120980
2007 Expense budget	$1,368,547.00
Cost per minute =	$1.22
Day rate	= $622.63 *for break even*

Cost + Profit	Day Rate	
Profit	120%	$747.16
	130%	$809.42
	140%	$871.69
	150%	$933.95
	160%	$996.21
	170%	$1,058.48
	180%	$1,120.74
	190%	$1,183.00
	200%	$1,245.27

Figure 10.7 Laboratory costing.

10.7 Test Risk

10.7.1 Type 1 Failures

A type 1 failure occurs when we begin to reject parts that are actually okay. That is why this situation is often called *producer's risk*. It is also known as an *alpha failure*. If we are dealing with safety-critical parts, some level of type 1 failure may be acceptable simply to assure ourselves that we are eliminating all failures.

10.7.2 Type 2 Failures

Type 2 failures are a little trickier than type 1 failures. A type 2 failure is when we pass a bad part as a good part. This situation is called *consumer's risk* because the issue will normally be discovered at the customer's site. We will only be able to put numbers on this failure insofar as the customer rigorously reports defects as they appear.

10.8 Calibration Issues

Actually, any pretense of meaningful testing will involve some level of periodic calibration of the instruments. Standards like ISO/TS 16949 absolutely require a calibration policy. We want to use calibration to reduce machine noise in the form of incorrect data and give a better picture of what we are measuring. This kind of calibration is similar to what we do before a machine capability study so we can assure ourselves that we are not investigating measurement noise.

In cases where a signal generator is not the primary measurement device, we can lengthen the calibration interval for that device. A typical example would be a transient generator that uses a calibrated oscilloscope to measure the signal. Because the oscilloscope is the final measurement device, it requires more frequent calibration. Of course, we will occasionally calibrate the transient signal generator so we have some assurance we are sending the correct signal to begin with.

In general, we favor more frequent over less frequent calibration. Although we expect a modest increase in the cost of maintaining the instruments, we also have some assurance that our devices are functioning as expected with no surprise delays that could have been avoided with some simple total productive maintenance.

Note

1. French, Simon, "Cynefin: Repeatability, Science and Values," *Newsletter of the European Working Group "Multicriteria Aid for Decision,"* Series 3(No. 17), 1, Spring 2008. Available at: http://www.inescc.pt/~ewgmcda/OpFrench.pdf (accessed June 19, 2010).

Chapter 11

Advanced Concepts

11.1 Test Impacts

11.1.1 Measurement Uncertainty

Often a consideration in calibration activities, measurement uncertainty describes an area about an observation (not an attribute) of some physical measurement that is likely to enclose the "true" value of that measurement. Measurement uncertainty can be modeled with a probability distribution, often the so-called normal distribution. Information for the calculations may come from two sources:

1. Actual statistical data collected by the metrologists
2. Documented values

Measurement uncertainty is related to systematic and random error of measurements and relies on the accuracy (the value of the mean) and the precision (the value of the variance) of the instrument used for the measurement. Intuitively, the poorer the mean and variance values, the larger the measurement uncertainty will be.

11.1.2 Factor Analysis

Factor analysis is a statistical approach we use to delineate variability among observed variables in terms of a lower number of unobserved variables called factors. In essence, we may have two or three observed variables that together embody another, unobserved variable. Factor analysis seeks out possible combinations. We represent the observed variables with linear combinations of the potential factors in addition to which we calculate some error terms. We may be able to use the information about interdependencies among and between observed variables to reduce the number of variables in the dataset.

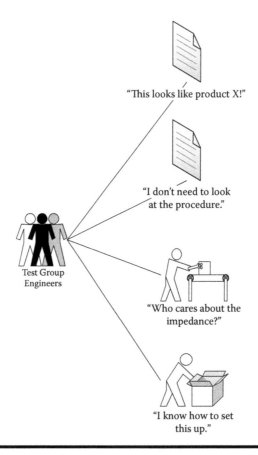

Figure 11.1   Test engineer bias.

11.1.3 Bias by Test Engineer

Figure 11.1 shows some simple ways a test engineer can sabotage a test suite.

11.1.4 "Passing"

During automotive testing (the type with which we are most familiar), we often hear project managers ask us if we have "passed," as if finding no anomalies during testing were a good thing. When using small sample sizes (i.e., we have *no* statistical validity), passing tells us absolutely nothing. Failure with such a small sample size tells us we either got very lucky or the product is probably terrible. We do not want to pass; we really want to demonstrate that the failure and destruction limits of the product significantly exceed the design limits, which are presumably based on customer considerations.

11.1.5 Sampling from Dependent Variable

Sampling from the dependent variable can lead to false inferences. A typical example of this kind of logical/data fallacy occurs when we look at a group of successful company executives and we research their backgrounds to see what is common. What is missing is the research based on a randomized sample taken from a population of executives. For all we know, the items in common may have nothing to do with their success, which is often the result of chance rather than volition.

11.1.6 "No Evidence of Cancer" or "Evidence of No Cancer"?

"Evidence of no cancer" means we have something that indicates the *absence* of cancer. In the case of "no evidence of cancer," we have nothing to show other than, perhaps, some level of diligence. We would prefer to have the first option. Any time we see the words *no evidence*, we know we are already dealing with a potential logical fallacy and a lack of understanding.

11.2 Construct Validity

11.2.1 Threats to Construct Validity

What is construct validity? This idea arises when we are concerned with whether our measurement system actually relates to our hypotheses, particularly when we are actually drawing inferences about the construct. Although construct validity concerns are more common in the social sciences, we think the idea also merits some concern with hardware and software testing.

11.2.1.1 Inadequate Explication of Constructs

Failure to adequately explain/delineate a construct may lead to wrong inferences about the relationship between operation and construct. Like many things on the planet, the more adequately we define our approach, the more likely it is that we will recognize that a situation is occurring.

11.2.1.2 Construct Confounding

Operations usually involve more than one construct and failure to describe all the constructs may result in incomplete construct inferences. Construct confounding is somewhat similar to confounding in designed experiments. In essence, we are not able to separate one construct from another, potentially leading to conflicting hypotheses.

11.2.1.3 Mono-Operation Bias

Any one implementation of a construct both underrepresents the construct of interest and measures irrelevant constructs, complicating inference. Mono-operation bias relates to the independent variable in our test suite—the mono-method bias relates to the results. If we merely use one version of a particular test or test sequence in a particular location at a particular time, we may not be grasping the full expanse of the purpose of the testing. Every implementation is defective relative to the construct (hypothesis usually in our case) from which it is derived. In other words, our test always has a particular slant to it regardless of our attempts to secure objectivity.

11.2.1.3.1 Mono-Method Bias

When all implementations use the same tactic (e.g., combinatorial testing), that tactic becomes part of the construct actually studied. *Mono-method bias* refers to our measures or effects, differentiating it from mono-operation bias. Our results may be only part of the story, particularly when we are using a hammer when we could also use a wrench.

11.2.1.3.2 Confounding Constructs with Levels of Constructs

If we observe no results from a specific test sequence we may not be stimulating the product to a sufficient level to detect even proper behavior, much less defective behavior. We must expand the range of our study to see where behaviors become detectable. We say we are testing, but we are barely challenging the product; hence, "testing" becomes a mislabeling issue.

11.2.1.3.3 Treatment Sensitive Factorial Structure

In some cases, our treatments (test runs) may affect the product sufficiently (modifying it) that we bias our results.

11.2.2 Translating Hypotheses into Test Scenarios

Herein lies the rub with construct validity. If we do not proceed with an open mind by bracketing the units under test—eliminating prejudicial approaches—we are effectively positing a construct that may not be valid for these parts. The goal here is to avoid the "I know, I know!" situation where test engineers look at an effect and *think* they know the cause.

11.2.3 Why Is This Important?

Our goal is the truth. Krishnamurti said the "truth is a pathless land." We approach each set of tests anew, doing our best to throw aside our prejudices, to achieve an unmediated appreciation of the behavior of the product. We suggest repeatedly in this book that the pragmatic test engineer open his or her mind to new ideas and different viewpoints. Any diagrammatic, team, or mental method that encourages a new look is a potential new weapon in the armamentarium of the test group.

11.2.3.1 Using Words Means Using Concepts

Virtually by definition, the use of words indicates that we are also using concepts. Clearly, careful definition and scoping are significant to understanding. We have seen descriptions like "the device locked up"; however, this could mean that the micro controller latched in a noncommunicative, nonfunctional state, or it could mean that the device software was "lost," or it could mean the device was unreactive under a saturated network scenario (much like denial-of-service attacks on the Internet).

11.3 Types of Bias

11.3.1 Causal Relationship

How do we know if cause and effect are related? In a classic analysis formalized by the nineteenth-century philosopher John Stuart Mill, a causal relationship exists if

1. The cause preceded the effect,
2. The cause was related to the effect, and
3. We can find no plausible alternative explanation for the effect other than the cause.

11.3.2 Causal Experimentation—Experimental Relationship

These three characteristics mirror what happens in experiments in which

1. We manipulate the presumed cause and observe an outcome afterward;
2. We see whether variation in the cause is related to variation in the effect; and
3. We use various methods during the experiment to reduce the plausibility of other explanations for the effect, along with ancillary methods to explore the plausibility of those we cannot rule out.

11.3.3 Validity

In general, we concern ourselves with four types of validity, which we see related in Figure 11.2.

11.3.3.1 Statistical Conclusion Validity

We can run into this situation when we disregard the effect of interactions (the co-variance). Sometimes, we do this deliberately when running screening tests designed

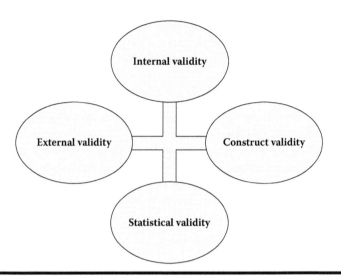

Figure 11.2   Types of validity.

to give us a reasonable idea of which factors are actually significant. That is why we recommend that the results of a designed experiment are always confirmed by subsequent testing to verify that our new hypothesis truly applies.

11.3.3.2 Internal Validity

Just because we can establish correlation does not mean that we have established causality. Covariance is not causation. We must always establish a mechanism for the behavior when working with correlation. In addition, it should be clear exactly how the mechanism functions and to what extent we are confounding main factors with interactions.

11.3.3.3 Construct Validity

Construct validity refers to the validity of inferences about the higher-order constructs that represent sampling particulars. We have discussed this idea somewhat already, so we already know that what we *think* about a test scenario, particularly the results, can have some effect on how we interpret what we see.

11.3.3.4 External Validity

External validity refers to the validity of inferences about whether the cause–effect relationship holds over variation in persons, settings, treatment variables, and measurement variables. This idea is particularly important when we compare bench testing or laboratory data against what we see coming from the field. We have conducted life tests on products that passed the required times with no failures and discovered later that these life tests were effectively irrelevant to what we actually saw from field use of the product.

11.3.4 Threats to Statistical Validity

11.3.4.1 Low Statistical Power

An insufficiently powered experiment may incorrectly conclude that the relationship between treatment and outcome is not significant. We control this situation in designed experiments when we pick an interval between the low and high values that is not representative of typical use. The other side also applies: if our interval is too small, we diminish the effect of the control factors.

11.3.4.2 Violated Assumptions of Statistical Tests

Violations of statistical test assumptions can lead to either overestimating or underestimating the size and significance of an effect. One common source of error is using too small a sample size to really learn anything about the product. This situation applies particularly when we are looking at attribute-based analyses.

11.3.4.2.1 Fishing and the Error Rate Problem

Fishing occurs when we go looking for specific relationships rather than allowing the relationships to reveal themselves through more objective analyses of factors and responses.

11.3.4.2.2 Unreliability of Measures

Measurement error can weaken the relationship between two variables and strengthen or weaken the relationships among three or more variables. We have seen designed experiments where the error was the most significant control factor! This situation can often mean that we are not

- Measuring the correct factor
- Adequately measuring the factor
- Understanding the effect of measurement uncertainty

11.3.4.3 Restriction of Range

Reduced range on a variable usually weakens the relationship between it and another variable. We see this occurrence with designed experiments when we do not use a reasonable range of values for the control factors.

11.3.4.4 Unreliability of Treatment Execution

We do not want to change a designed experiment in the middle of executing the "recipe" for the treatments. In some cases, we can recover from this poor protocol with a split design, but often it becomes impossible to recover from impromptu changes. When we cannot recover, we just wasted time and money.

11.3.4.5 Extraneous Variability in the Test Setting

We need to either account for extraneous variability as noise or provide some kind of designed block to remove this variability from contributing to a complete misunderstanding of our experimental/test results.

11.3.4.6 Heterogeneity of Units

If our test samples are wildly different in behavior, our results are unlikely to have much validity. We need to understand the process that produced these samples. We see this situation when we are using prototypes in the laboratory, as prototypes are often hand-built rather than following the normal production process.

11.3.4.7 Inaccurate Effect Size Estimation

Some statistics systematically overestimate or underestimate the size of an effect. That is why we have corrective statistics such as the Bonferroni correction (used to maintain the family-wise error rate). Some statistics such as t-statistics may only become robust with relatively large sample sizes, even though that particular method was originally designed for use with specific, and often small, sample sizes.

11.3.5 Threats to Internal Validity

11.3.5.1 Ambiguous Temporal Precedence

Lack of clarity about which variable occurred first may yield confusion about which variable is the cause and which is the effect. This situation is a violation of the Mill rules and can lead to significant confusion. The philosopher Nietzsche would occasionally point out cases where he thought cause and effect were reversed.[1]

11.3.5.2 Selection

If we have systematic differences in response variable characteristics, we may see an effect. Essentially, this kind of situation happens when our samples are not truly random, as is often the case in design/manufacturing facilities when we are dealing with small prototype lots or the quality engineers pick the first or last parts of the build.

11.3.5.3 History

Events occurring simultaneously with testing might cause the observed response. Basically, we are looking at potential coincidences with this bias. Usually, we reduce this possibility using careful replications of the original treatment. In El Paso in 2006, for example, we had unusually high quantities of rain in August (a year's worth in 1 month!), and we had to purchase a dehumidifier for some of the electronics tests in order to keep our relative humidity measurement within the required standard.

11.3.5.4 Maturation

In testing situations where we have already "beaten up" the product, we may see degradation effects that are not realistic. When this occurs, we may be victims of irrelevant failure modes. We will, of course, see maturation effects with products that sustain a long field history. Sometimes, we may find difficulties separating routine degradation from special causes.

11.3.5.5 Regression

During regression analyses, we should be careful when dealing with the effect that outlying values can have on the analysis of the response. Legitimate techniques exist for eliminating outliers: the method of hinges, the quartile approach, and the Grubbs test.

11.3.5.6 Attrition

If we begin to lose sample parts, we may reduce the randomness of our test part lot, thereby biasing the analytical results. These effects become more significant if the part loss is related to test conditions, especially environmental conditions. For example, we may have parts dropping out due to dendritic shorts caused by the condensation of moisture on contaminated printed circuit boards.

11.3.5.7 Testing

We need to be careful when retesting parts because the results of a given test can influence our understanding of subsequent runs of the same test. We need to ensure that we understand the effect of this approach on the experimental response.

11.3.6 Threats to External Validity

11.3.6.1 Interaction of the Causal Relationship with Units

An effect found with specific kinds of units might not hold if other kinds of units had been studied. The question is one of generalization. How far can we generalize from one set of sample pieces to another? Our test reports say that the results only represent the samples actually tested, which functions effectively as a disclaimer from unwarranted extrapolation.

11.3.6.2 Interaction of the Causal Relationship over Treatment Variations

A response found with one treatment variation may not apply to other variations of that treatment, when that treatment is combined with other treatments, or when only part of that treatment is used. This potential bias can usually be counteracted with appropriate replication of our experiments. In our experience, automotive testing is usually severely remiss at performing the correct number of replications.

11.3.6.3 Interaction of the Causal Relationship with Outcomes

An effect found on one kind of response observation may not hold if we used other response observations. In essence, we need to use our measurement system consistently or we will clearly achieve inconsistent results. We usually use the same test engineer for the duration of a specific test.

11.3.6.4 Interaction of the Causal Relationship with Settings

An effect found in one kind of environment may not hold if we use other kinds of environments. This occurrence is the curse of bench testing. In fact, many laboratory tests require a "mild" environment as part of standardizing the test. Consequently, we think it is short-sighted to conduct all product validation tests on the bench.

11.3.6.5 Context-Dependent Mediation

An explanatory mediator of a causal relationship in one context may not mediate in another context. In this situation, if A depends on B to get to C (a transitive action) under a specific environment setting D1, then B may not affect result C in another environment setting D2. In one case we observed the supplier did not understand the rigors of the environment in which a product was to be used—lakes and seashores—which led to a nonworking and leaky product. The required tests were conducted in the bench environment and the actual product environment was significantly harsher than that.

11.3.7 In Specifying and Selecting the Study Sample

11.3.7.1 Popularity Bias

With some automotive warranty issues, "word gets around." Such an occurrence can provide opportunities for unscrupulous vehicle owners to receive new parts. Guess what? Your part just became *popular.*

11.3.7.2 Referral Filter Bias

Because the warranty group in a company/laboratory sees the units that have grossly misbehaved, they may get a mistaken idea of the frequency of events that are truly rare. We have found it critical to compile good field data to counteract or inoculate ourselves against this effect. This particular bias can be yet another example of the "I know, I know" behavior we have seen in product review meetings.

11.3.7.3 Diagnostic Access Bias

The customer and the supplier may use different names for specific failure modes. Additionally, the tool sets used by the two parties are often different. If the customer is mammoth and the supplier is petite, we can see significant differences in test capability and what each party can "see."

11.3.7.4 Diagnostic Suspicion Bias

If we know that a product has been exposed to specific environments, it may influence our diagnosis of the putative causes of the observed failure mode. One example occurs when we have demonstrated that temperature is a significant factor in the laboratory and we then go looking for temperature effects in the warranty data.

11.3.7.5 Unmasking (Detection Signal) Bias

An innocent environmental exposure may become suspect if, rather than causing a failure mode, it causes an effect that precipitates a search for the failure mode.

11.3.7.6 Mimicry Bias

An innocent environmental exposure may become suspect if, rather than causing a failure mode, it causes a benign failure bearing some resemblance to the original failure mode of interest.

11.3.7.7 Previous Opinion Bias

The tactics and results of a previous diagnostic process on a patient, if known, may affect the tactics and results of a subsequent diagnostic process on the same patient. The same situation can occur with hardware and software testing when we begin to develop a "favorite" set of tests for a specific issue.

11.3.7.8 Wrong Sample Size Bias

Samples that are too small can prove nothing; samples that are too large can prove anything.[2]

11.3.7.9 Selection or Berkson Bias

If warranty rates differ for different product families, the relation between usage exposure and failure becomes distorted in our analyses. Basically, we have a systematic error caused by nonrandom sampling. If we base our warranty analysis on only the failed parts, we are really looking at only a subset of the population of parts. Additionally, if the selection is nonrandom, we may emphasize one failure over other potentially more significant failure modes.

11.3.7.10 Prevalence-Incidence (Neyman) Bias

If we see a coincidental factor in our analysis of failure causation, we may include that factor as a part of the transfer function when, in fact, the factor contributes nothing at all to the response.

11.3.7.11 Diagnostic Vogue Bias

If we see a frequent effect and we have recent and standard "cause" for that effect, the tendency is to assign that specific cause to that effect. We must re-emphasize: It is a logical fallacy called "affirming the consequent" when we try to reason from effects to causes.

11.3.7.12 Missing Data Bias

Missing data may be missing because it is normal, negative, never measured, or measured but never recorded. The integrity of the measurement system is paramount. Automotive suppliers generally check their measurement systems using gage reproducibility and repeatability studies, machine capability studies, and other measurement system checks, including extensive requirements for calibration.

11.3.7.13 Membership Bias

Membership in a group (rental vehicles) may differ systematically from that of the general population of vehicles because of the different care received by rental vehicles. We have seen this situation where one provider of rental medium-duty trucks had significantly higher failure rates than that of the general population.

11.3.8 In Measuring Exposures and Outcomes

11.3.8.1 Insensitive Measure Bias

When outcome measures are incapable of detecting clinically significant changes or differences, type 2 errors occur.

11.3.8.2 Underlying Cause Bias (Rumination Bias)

Engineers can ponder about possible causes for product failure and, hence, exhibit some level of bias toward their analysis of the parts. Here again, we see the "I know, I know" phenomenon. Please understand that this phenomenon can be subtle—usually nobody is jumping up going "I know, I know" in actual fact but simply and quietly exhibiting the bias.

11.3.8.3 End-Digit Preference Bias

In converting analog to digital data, observers may record some terminal digits with an unusual frequency—sometimes attributable to numerical round-off decisions. A search of the literature suggests that this bias is considered significant when making judgments based on blood tests, among other medical results.

11.3.8.4 Unacceptability Bias

Measurements that hurt, embarrass, or insult customers may be systematically refused or evaded. We have seen cases where the results were modified with different terminology to "soften" the result for the customer and to make it less obvious that an issue had occurred during testing. We demand that our tests always, at a minimum, indicate an observation when they see any item that seems amiss, even if it later turns out to be irrelevant.

11.3.8.5 Expectation Bias

Observers may systematically err in measuring and recording observations so that they agree with the expectations of the tester. And, yes, this bias can be another example of "I know, I know." Expectations of specific results can be insidious to the quality of the report and, more importantly, to the conclusions derived from the results.

11.3.8.6 Substitution Bias

The substitution of a control or noise factor that has not been established as causal for its associated outcome can invalidate a set of results. A more correct approach here would be to run a new set of treatments with the new factor or factors.

11.3.8.7 Family Information Bias

Although this bias is generally seen in medical situations, it can also affect analyses because we are often dealing with a family of products. If a family of products acquires an unsavory *reputation*, we can see the test engineers anticipating failures as well as coloring the analyses of failures they do observe.

11.3.8.8 Exposure Suspicion Bias

Knowledge of the failure mode of the part may influence both the intensity and outcome of a search for exposure to the putative cause. In short, we believe we already know what the issue really is. We see this in meetings where somebody clouds the minds of the team with a passionate "I know, I know" and proceeds to explain all about the failure mode and its cause. We need to proceed during an investigation using something akin to the phenomenological reduction, where we "bracket" the issue and examine on the basis of its ments, rather than some prejudgment about cause and effect.

11.3.8.9 Instrument Bias

Defects in the calibration or maintenance of measurement instruments may lead to systematic deviation from true values. Instrumentation should be on a meaningful recalibration interval to avoid this kind of problem. Tools with moving parts, such as automated screwdrivers in a factory, need to be recalibrated frequently to avoid producing out-of-specification products.

11.3.9 In Analyzing the Data

11.3.9.1 Data Dredging Bias

When data are reviewed for all possible associations without prior hypotheses, the results are suitable for hypothesis-forming activities only. We are shooting at the wall and drawing a target around the bullet holes! We should not assume that a target exists, and we should certainly not create one after the analyses. We know that the hypothesis must be proposed *before* the test and not after it. We are always testing a hypothesis rather than searching for one.

11.3.9.2 Tidying Up the Bias

The exclusion of outliers or other untidy results cannot be justified on statistical grounds and may lead to bias. The removal of outliers should always be accompanied by a rationale for doing so. Sometimes it is the outliers that are truly important during a test. Furthermore, the original results for any test or suite of tests should always be available for subsequent review by an objective examiner. We should never present tidied results without indicating the method for tidying up the data and, perhaps, presentation of what the original data looked like—often, a well-constructed graph is sufficient to make the rationale for the tidying obvious to all but the most jaundiced observer.

11.3.9.3 Repeated Peeks Bias

Repeated peeks at accumulating data in a randomized trial are not dependent and may lead to inappropriate termination. In sort, we "jump the gun" because we believe that we already have sufficient data to make a conclusion. All experiments should meet the complete recipe to ensure that we have some reasonable basis for our conclusions and subsequent decisions.

11.3.10 In Interpreting the Analysis

11.3.10.1 Cognitive Dissonance Bias

The *belief* in a given mechanism may increase rather than decrease in the face of contradictory evidence. We should not be surprised at this point to see another example of the "I know, I know" bias again. We have seen some cases where even the experimental technique itself was a form of bias because the testers thought they already knew what was occurring in a production process. Using experimental screening tests is one way to objectify our approach, particularly if the screen is large enough to look at all reasonable factors and even some unreasonable ones.

11.3.10.2 Magnitude Bias

In interpreting a finding, the selection of a measurement scale may affect the interpretation of the results. The situation may become worse when we are dealing with off-scale results. For example, we need to understand whether we are using interval or ratio scales (Celsius versus Kelvin with temperature) when we make our calculations. Arrhenius equations either use the Kelvin scale directly or have a conversion factor built into the formula to achieve the same result with Celsius temperatures.

11.3.10.3 Significance Bias

The confusion of statistical significance, on the one hand, with severity, on the other hand, can lead to wasted time and meaningless conclusions. Statistical significance is the complement of confidence. It may yield an idea of the power of our result but it has nothing to do with the severity or criticality of the results.

11.3.10.4 Correlation Bias

Equating correlation with causation leads to errors of both kinds. Causation is not always easy to prove; and even with Mill's criteria, the results can be challenged by someone influenced by the philosophy of David Hume, the great skeptic. In common parlance, we hear the term *guilty by association*, which is a simple way of indicating Hume's argument—he maintained that what we called cause and effect was an inference based on temporal association.

11.4 Reliability and Confidence

11.4.1 Test Results Reliability

Reliability of test results relates to our ability to replicate results. Some tests, such as bulk current injection, are highly reliable and can be performed repeatedly with relatively similar results. Other tests, such as radiated susceptibility, are extremely sensitive to several factors such as antenna position, sample position, and the position of device measuring the field intensity. This pattern of behavior suggests that is it wise to perform objective tests to characterize the degree to which we can replicate results. We would concern ourselves with sample size, sampling error, homoscedasticity, and any possibility we have of finding another objective indication of the quality of our results.

11.4.1.1 Quantities

In general, the law of large numbers will apply to calculations of reliability unless it is compromised by either lack of randomness (a concept that is difficult to execute) or heteroscedasticity (clumped data). We think it wise to always plot the data if it is possible to do so and let the human eye do a quick estimate of both randomness and heteroscedasticity. We can use other analytical techniques later if the "eye check" suggests that such is merited.

11.4.1.2 Duration

Time-based testing is always an issue, with real field data being the acid test for duration testing. At best, we can hope that we are presenting an environment of reasonable verisimilitude or that we are doing an elephant's foot test that shows the robustness of the product beyond a doubt.

11.4.2 Test Results Confidence

11.4.2.1 Quantities

The law of large numbers suggests that we increase our confidence when we increase the sample size as long as the samples are truly random samples. We need to be on the alert for heteroscedasticity to avoid issues with clumping of data. Most simple confidence calculations have an inbuilt assumption of homoscedasticity that can be challenged.

11.4.2.2 Duration

We see the same problems with time-based testing in regard to statistical confidence that we do with statistically based reliability.

11.5 Life Testing

Guessing, extrapolating, or intuiting (all really just guessing by other names) the environment in which a product will be subjected to does not go far in securing the success of the product. Acquiring this information is frequently not a trivial task. The consequences of missing this information cause a variety of negative impacts and costs. Acquisition of this information allows the test designer to be able to plan tests that will replicate this *life* exposure, even compressing the time to achieving this exposure. All of this gives the organization a good understanding of the capabilities of the product when pressed into the service for which it was designed.

Underestimation of the environmental factors has an impact on new products, resulting in premature failures, high warranty dollars, high customer dissatisfaction, and the opportunity costs for lost projects. These faults can influence customer perception of the product and, if severe enough, possibly lead to product recalls and legal action taken against the supplier, a situation in which everybody loses. If the product meets the customer specification but fails in the field, then the customer organization will be the organization bearing the brunt of the consequences of the failure. This is still a situation in which everybody loses.

Even if the problem is not a "hard" failure, functional performance anomalies in ambient environments unpredicted during the development effort can have an adverse effect on the customer. For example, the key product characteristic of a tire pressure monitoring system is the transmission and reception of the data via wireless at 400 MHz. At elevated temperatures, the receiver's center frequency drifts. The transmitter is intermittently received, and the customer perceives the product to have intermittent operation. Understanding the ambient environment impact on the parameters that make the product meet the demands of a world of unexpected stresses makes for a better result.

Overestimation of the demands on the product also adds to the cost. An excessively hardened product will have increased material costs. This situation presents an opportunity cost for the organization: excessive investment needed for the product affects the number of projects or products that may be launched. Additionally, the cost of the product either increases to maintain the same profit margins, or the cost does not increase and the margins on the product decrease. Insufficient diligence may reveal that the business case for the product was poor from the start.

11.5.1 Determine Stimuli the Product Will Be Subjected to in Real Life (Including Frequency and Duration)

For products that are in an easily predictable environment such as a desktop PC, this work is relatively easy. Similarly, it is true for products that are used in a typical manufacturing setting. Measurements are taken of the area in which the product will be functioning. These measurements of the ambient environment are used to determine the probable operating environment of the product.

In the automotive world, for example, ambient thermal stimuli of the product are typically understood by taking measurements of summer test and winter test vehicles. These measurements provide the absolute maximum temperature elevation experienced by the product during the test vehicle runs.

One of the first things is to correlate the real world and the testing needs. In the heavy truck world, in order to determine product use, customer vehicles are set up with instruments to capture key information about the use of the proposed product. This is done to understand product time quantum. A set of vehicles is monitored for a period of time to assess the use of a particular function. For example, we might find that in 3 months, the function is used 12 times, during 40,000 miles. It can be projected that the function will be used 48 times in 120,000 miles (expected truck miles in 1 year and assuming linear extrapolation). The expected life of the vehicle is 1,000,000 miles; the product activation could be extrapolated to be 400 times, which constitutes the total use of the functionality over the life of the vehicle. Some buffer factor may be added to this to represent the most used case.

So how do you identify the key external demands on the product? Knowledge of the weakness of the constituent parts is important. Thermal demands on electrical/electronic components significantly affect performance and longevity as electronic components display degradation that can be modeled with an Arrhenius equation. Past product development experience is important in understanding the weaknesses of the product, and field performance data is useful also. Field performance data from a similar product will allow for statistical analysis of failure rates and different failure modes. If the development firm is desperate for some kind of assessment, they can also use an approach called the *parts count method* for determining reliability. The parts count approach derives from the MIL-HDBK-217F and uses estimates of life for families of electronic components under a variety of specified conditions. The downside of this approach is that the values determined from the handbook are usually extremely conservative, meaning that the part may be over-engineered for reliability if the parts count approach is the only tool used to estimate the reliability.

This same approach that we used to understand the operation of a function or product can be exercised to gain some understanding of the environment in which the proposed product must work. Several locations representing the area where the product might be used are set up with instruments to learn about the variety of stimuli to which the product is subjected during use. In this case, we monitor the ambient environment of the proposed product mounting position. We could measure thermal exposure, voltage transients, vibration, or other environmental information suspected to be critical.

11.5.2 Apply Multiple Stimuli Simultaneously

In actuality, the product will not experience a single external environment stimulus at one time in isolation from other stimuli. The product will receive vibration at low temperature, high humidity, elevated temperature, and so on. These combined interactions can cause performance issues from tolerances as well as outright failures of the product. In some cases, the interactions will be more detrimental than any single environmental factor by itself. Information gained from the stimuli assessment is used to generate key environmental combinations witnessed or logged in the field.

The answer is that just about any combination can be present in the field, so adequate testing can be difficult to achieve. In some cases, the most critical variable (or "factor") is time. This is especially true when an undesired event is particularly time dependent. How long is long enough? Basically, this question is unanswerable without substantial information. If we already have that information, the issue may have already been designed out of the product. Another type of time-dependent issue that can cause substantial heartburn is the intermittent failure and the means by which it is generated. Test standards typically do not provide much guidance for testing that will discover these types of failure. Oftentimes, a form of stochastic testing will elicit such a failure. Stochastic testing occurs when we set up our test stimuli such that they present randomly both in order and in duration. The downside of some stochastic testing is that it may be unrepeatable. These kinds of tests can become repeatable if they are scripted; however, the element of truly aperiodic behavior may diminish.

The choice of stimulus or stimuli can also be based on the desired information we wish to elicit from the product. For example, we can use a mechanical shock test on electronic samples (printed circuit board products) to test the integrity of the solder joints. In this case, the verisimilitude of the experiment is not identical to that in the field, but we can extract information from the sample parts quickly, comparing an older design with a newer design with a minimum of sample pieces. If the newer design is not demonstrably better than the old, we would not pursue further testing with more complicated multi-environmental overstresses; in short, we have saved both money and time.

11.5.3 Sometimes Hard to Correlate with Other Tests

Multi-environmental tests can be difficult to correlate with other multi-environmental tests unless the experimental conditions are nearly identical. One of the reasons test customers often choose one-factor tests is for precisely this reason: One-factor tests are much easier to control and to compare.

11.5.4 Highly Accelerated Life Testing

This type of testing is associated with driving the product's reliability by placing unusual stresses on the units under test. Highly accelerated life testing (HALT) often consists of multiple environmental stresses. These stresses increase over time and subsequent failures are critiqued to determine the cause and design corrective actions. The ever-increasing stresses go to the limits of the material, so ultimately failures will be witnessed.

HALT testing cannot be used to calculate an anticipated life for the product—at least, not directly. If we use B vs. C testing and test a new product against an older product with field history, we can sometimes say that the new product is more robust than the older product (if it outlasts the older product!) and, hence, we anticipate, at a minimum, superior life to that seen in the older product. This approach also allows us to use the Tukey quick test (end count test), a nonparametric calculation that can yield very nice confidences if the products are significantly different in robustness.

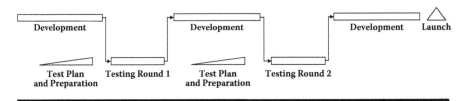

Figure 11.3 Testing integrated into the development.

11.6 Testing Team Early Involvement

While many have the perspective that testing is the last major task in product development, this is actually not true—if we want to deliver the product on time and at the desired launch quality. An accumulation of schedule and budget slips during the development effort frequently slips the final delivery schedule. When this happens—and we have not been testing all along—we find problems that will most assuredly delay the product launch or will require launch rescheduling. There is no maneuvering room for the required testing and there will definitely be significant amounts of uncertainty regarding the quality of the launch. If we had been testing all along, it would be possible to make some rough predictions based on previous cycles of testing. Figure 11.3 shows a typical phasing of testing versus development. Generally, the earlier we can test, the better off we will be because we give the development team some "wiggle" room for changes to the product.

Additionally, when we bring the test team in at the end, the test team only starts to gain experience with the product at the eleventh hour. This behavior is unlikely to secure a successful launch. The team must understand the product, develop test cases, and build fixtures in order to effectively verify the product. Much of the required information is in the product specifications (which can be elaborate) and, if the product is a collection of products (a system), the learning required to understand the product and develop reasonably representative test suites may not occur quickly enough. Forcing the testing team to crash the schedule because other parts of the development organization did not properly understand the nonlinearities in the development process is really not a very good solution. A more rational solution to this issue is to delay the launch until some previously designated criteria for success have been met.

An offshoot of crashing the test schedule is that the test engineers become tired due to the unusual overtime hours and can begin to make mistakes. In the commercial vehicle business (as well as passenger vehicles), the stakes are too high to ignore the physical and mental condition of our test group. The goal is not to check off some abstractions on a checklist but to determine the behavior of the product—to characterize it completely so we have some level of confidence that we are not delivering a deadly weapon to the end user.

11.6.1 Enables Past "Finds"in Past Testing to Be Considered during the New Design

Not only do the test teams need time to understand the product they will test, but they will find a benefit in tapping into past product failures. When the test people are included in the development activities, legacy design practices may leave some traces in the current product under development that can be uncovered and available for the test team to use to their advantage by incorporating historical information into their plans in the form of "lessons learned." Rather than wait to see a failure that the organization has seen before, product development "historians" can bring these potential issues to the development group for their critique.

This approach can also be instantiated by including lessons learned in the design failure mode and deffects analysis (DFMEA) and the process failure mode and effects analysis (PFMEA). However, this method will work only if the test team and the development team spend the time to review this critical information. Certainly, items that yielded high severities or criticalities should be reviewed always.

In a sense, the use of legacy information is similar to reusing design blocks in software and hardware development. We can save substantial time by using information from the past as long as we are careful to avoid introducing any of the biases we discuss in this book. In short, we should remain alert for any chance for reuse of existing material!

11.6.2 Early Development of Test Plan and Review Ensures Plan Suitability

An additional benefit of integrating the testing into the development process is the ability to review the test plan as early as possible. Reviewing the plan makes it possible to make alterations to the scope of the verification activities; in short, we have time to maneuver our test plans to achieve the best results. With enough early work, it becomes possible to make intelligent decisions about areas of risk for the product or service and to emphasize known high-risk areas in the test documentation. As we have discussed elsewhere, if we have a product simulator, we can begin testing as soon as the simulator is operational. This approach provides an opportunity to validate the simulator and the test plans simultaneously, as well as providing early warning of potential issues.

11.6.3 Provides Time to Create Any Specialized Tools or Fixtures for the Testing

To test efficiently, we need to develop a test fixture or, in some cases, we need to automate the test suite. These special tools can often take substantial time to develop—sometimes half a year or more. Time is required to do the work and, even before then, to determine the best way to do the work. The test group must understand the product and organizational goals and develop test methods to characterize the product. When automation is possible and the organizational philosophy is consistent, we may need to spend the time developing test scripts, which is software development by another name. If it is determined that a thermal and vibration profile should be performed simultaneously, then we need the time to develop the fixtures or outsource the activity. As we mentioned elsewhere, simulation can help us perform early testing, although it may not provide much information about a hardware fixture.

Notes

1. Nietzsche, Friedrich, *Twilight of the Idols* (London, UK: Penguin Books, 1968), p. 58.
2. Sackett, David L., "Bias in Analytic Research," *Journal of Chronic Disease*, 32(1-2), 61, 1979.

Chapter 12

Software Test Documentation

Software test documentation fulfills the same roles as the hardware test documentation. First, this documentation must clarify the scope of the testing to be conducted. We follow the scope by instantiating the philosophy of the testing, focusing on the test demands, planning the approach, and subsequently the details of the test needs.

12.1 IEEE 829

IEEE 829-1998, Standard for Software Test Documentation[1]:

1. Test plan identifier
2. Introduction
3. Test items
4. Features to be tested
5. Features to not be tested
6. Approach
7. Item pass/fail criteria
8. Suspension criteria and resumption requirements
9. Test deliverables
10. Testing tasks
11. Environmental needs
12. Responsibilities
13. Staffing and training needs
14. Schedule
15. Risks and contingencies
16. Approvals

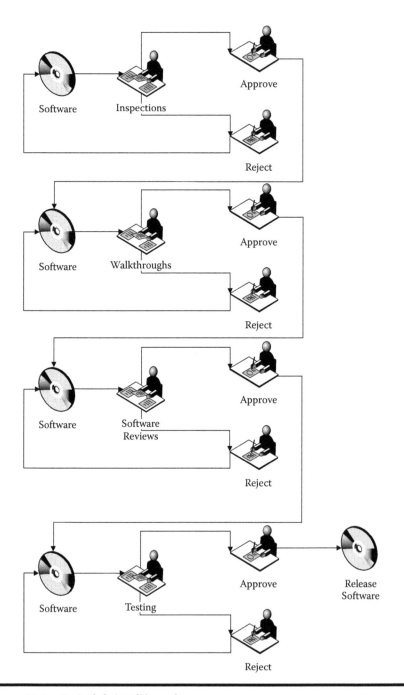

Figure 12.1 Typical defect life cycle.

12.2 Defect Life Cycle

Figure 12.1 shows the defect life cycle for one team. Notice the iterations that occur because the software is rejected for quality defects. What we are seeing in Figure 12.1 is really a test-analyze-and-fix product development process.

Note

1. Institute of Electrical and Electronics Engineers, Inc. *Standards for Software Test Documentation* (New York: IEEE, 1999), p. 3.

Chapter 13

Configuration Management

A fundamental process of product development is configuration management. Configuration management systems help to deliver coordinated software and hardware. They do this by easily identifying features and functional content within a particular software and hardware availability. In fact, it can be said, "You can't effectively test if you do not know what you have available to test." In complex system development where there are multiple hardware and software revisions, and the lack of a configuration management system can cause test failures due to combinations of software and hardware that were never designed to work together or verification of functions or features that are not yet available. We know from experience in various companies that an optimally running configuration management program will reduce erroneous fault reports.

Not only does configuration management aid in testing, but we also can have clear traceability of function and feature growth over specification revisions rippling through hardware and software deliveries and ultimately into the testing aspects of the project. Throughout the process, we can always

- Clearly identify design changes and subsequently test changes
- Identify compatible components to construct a functioning system
- Execute prerelease containment, making it possible to *know* the impact of hardware and software changes

Additionally, the tests conducted will only exercise those features that are available, thereby reducing testing of portions of the system or functions that are not even available. This approach streamlines test times, only testing what is available for test. Manufacturing firms with good configuration management systems bring their part

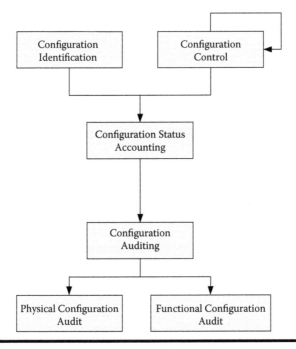

Figure 13.1   Overview of configuration management process.

numbers and product releases under control, which will enhance our ability to know what features we will be testing.

An effective development enterprise depends completely on robust configuration management, whether the product is software, firmware, hardware, or all three. Sending reports of failure to a customer when the problem is a compatibility issue with the constituent parts does not inspire confidence. Figure 13.1 is a high-level overview of the configuration management process. Configuration control loops back on itself because the control portion is definitely a repetitive process. Of course, any modification to the product would require another voyage through the total process to ensure that we met all of the requirements for change management.

13.1 Configuration Management Areas

Configuration management should be applied to design and test specifications (documentation), models, simulation, hardware, software, the combination of hardware and software, and system integration. The truth is that configuration management should be applied to just about anything where uncontrolled or unaccounted-for change has a negative impact on the product and delivery. We show a version of a software configuration management system in Figure 13.2.

■ Product baseline—allows comparison of the original customer needs to changes (scope)

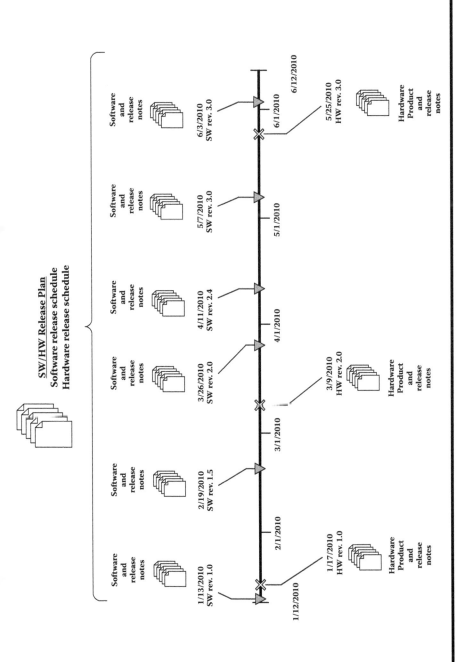

Figure 13.2   Example of configuration delivery plan.

- Product change management—controls the change introductions, reduces chaos
- Number of product changes—change happens, and change impacts cost (money) and deliver (time)
- Test coverage—identified in test configurations
- Configuration audits—*trust but verify* the configuration management plan and the deliveries of hardware and software match

13.2 Planning

It is necessary that we plan and coordinate the configuration of the software and the hardware deliveries (functional and performance content) to both testing and the customer. In short, functionality, deliveries, testing, and production follow a master plan rather than pandemonium. With appropriate planning we will always know the following items:

- Features and function levels within a specific revision:
  - Example: Instrument Cluster, On-Board Diagnostics (OBD) lamp rev. 1.0
- Software revision level
- Hardware revision level
- Subassembly revision level
- System integration (staged introduction of iterations of the subsystem)

In essence, we should have a grip on most of the 18 items in a production part approval process package. IEEE Standard 1220 favors the master plan approach to system engineering, providing an alternative to standard project management practices.

13.3 Elements of Configuration Management

According to Military Standard (MIL-STD) 973 and Military Handbook (MIL-HDBK) 61, there are four requirements for a configuration management system. These four items form the framework on which we hang the remainder of the configuration management system (Figure 13.3).

1. Configuration identification
2. Configuration control
3. Configuration status (accounting)
4. Configuration authentication (auditing)

The layout of a very complete configuration management plan might use the following approach:

1. Introduction
   a. Purpose
   b. Scope

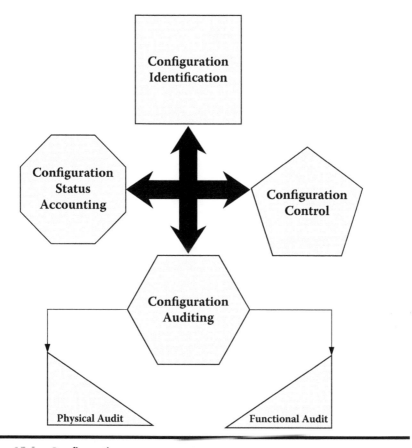

Figure 13.3 **Configuration management components.**

    c. Approach
    d. System Overview
    e. Project-Defined Configuration Items
    f. Document Overview
    g. Terms and Definitions
2. Referenced Documents
    a. Customer Documents
    b. Internal Documents
    c. Customer Standards
    d. Internal Standards
    e. National Standards
    f. International Standards
3. Management
    a. Organization
    b. Boards

    c. Resources
       i. Plans, Schedule, and Budget
       ii. Facilities
       iii. People
       iv. Configuration Management System
    d. Risk Management
4. Phases and Milestones of Project (if project-based system)
    a. Concept Phase
    b. Development Phase
    c. Production Phase
    d. Utilization Phase
    e. Support Phase
    f. Retirement Phase
5. Data Management
    a. Digital Data Identification
    b. Data Status Level Management
    c. Maintenance of Data and Product Configuration Relationships
    d. Data Version Control
    e. Digital Data Transmittal
    f. Data Access Control
    g. Status Reporting
    h. Data Security and Classification Management
    i. Maintenance of Disaster Recovery Data
6. Configuration Identification
    a. Selection of Configuration Items
    b. Formal Baseline Establishment
    c. Identification Methods
    d. Developmental Configuration–Corrective Action Process
    e. Configuration Management Libraries
7. Interface Management
    a. Interface Requirements
    b. Interface Control Working Group
    c. Interface Control Document (ICD)
    d. Block Diagrams
    e. P-Diagrams
8. Configuration Control
    a. Boards
    b. Baseline Change Process
9. Configuration Status Accounting
    a. Records
    b. Reports
    c. Requests for CSA Reports
10. Configuration Audits
    a. Functional Configuration Audit
    b. Physical Configuration Audit
    c. Audits and Reviews of Configuration Management Program

13.3.1 Configuration Identification

Establishing configuration identification requires the identification of all configuration items and a rational nomenclature policy. In Department of Defense software implementations, these are known as computer system configuration items (CSCIs or "siskies"). Generally, we pick an item and define the initial version of it as a baseline. This baseline will function very much as a baseline in a project management tool; namely, it allows us to know where we started and to detect and record change as the project or product development proceeds.

With software, we usually are well served using dedicated software configuration management tools; for example,

■ Subversion
■ Revision control system (RCS)
■ Concurrent version system (CVS)

All these programs lock out other users from making changes during an edit and provide some level of reporting. They basically mechanize the software build process by defining what modules and revision of the modules are included. Most "make" programs have a means of accessing the software configuration management system automatically.

When we have a combination of hardware and software, known as *firmware*, we need to have coordination with these embedded projects. The situation is set up as an assembly part number with two subassemblies: the chip and the software—together they make the subassembly. For example,

■ Subassembly
■ Product
■ Firmware

Hardware configuration management is accomplished through the part number system and the associated bill of materials. The part number may include a revision level in the designator or may be an individual part number with each new iteration. The structure of the bill of materials is unimportant; however, it is important to know that no matter the structure, the bill of materials should also be under configuration management control.

In many cases, we will define a part number, describe the feature content, apply a specific function and revision level, and we may even define a further level of refinement. At every stage, configuration identification provides a process and product that is documentable and controllable.

13.3.2 Configuration Control

Configuration control is the part of configuration management where we supervise changes to the configuration of the product, whether it is software or hardware. In some cases, we may re-baseline the product. In general, we usually have only one version in release at a time. Control can become complicated during transitions when more than one version may be available; in these cases, we can control with part numbers. Management can also become complicated with multiple component configurations for a system, particularly in the age of mass customization of products. With motor vehicles, we often see attribute-based requirement or line sequencing, in which the parts are customized to the vehicle identification number (itself a form of configuration identification). This production complication has ramifications to the verification activities. Specifically, these production variations must be accounted for in the testing activities.

Typical configuration control baselines might be the following:

- Allocated baseline (ABL)—systems-level allocation
- Functional baseline (FBL)—functional allocation also used for verification
- Product baseline (PBL)—includes acceptance testing and hardware and software

Regardless of the baseline, the identified configuration involves an agreed-upon description of the attributes of a product by the development team. Once agreement is reached, this accord constitutes the starting point for any future change (see Figure 13.4). There should be an approved and released document or documents (revision) for every specific version, the purpose of which is to provide a solid foundation for managing change. These baselines always include the currently approved and released configuration documentation and, with software, a released set of files making up a software version and related configuration documentation.

13.3.3 Configuration Tracking (Status Accounting)

We use configuration status accounting (CSA) primarily for the reporting feature, although its significance is much broader. Under CSA, we record changes and update configurations when items change and we issue reports. When dealing with software, we might audit the frequency of check-in/check-out to verify that the developers are protecting our intellectual property. We may audit the release documentation to confirm that the proposed or planned configuration has been delivered. Additionally, we might take a look at the count of changes to get an idea of the stability of a given release. An exceptional amount of churning may indicate a high level of change and, in some cases, may be indicative of poor hardware or software.

CSA can be supported by software (e.g., the open source product Subversion), and it is definitely a component of product data management software, product life-cycle management software, and document control packages. However, it is also possible to use ubiquitous spreadsheet software.

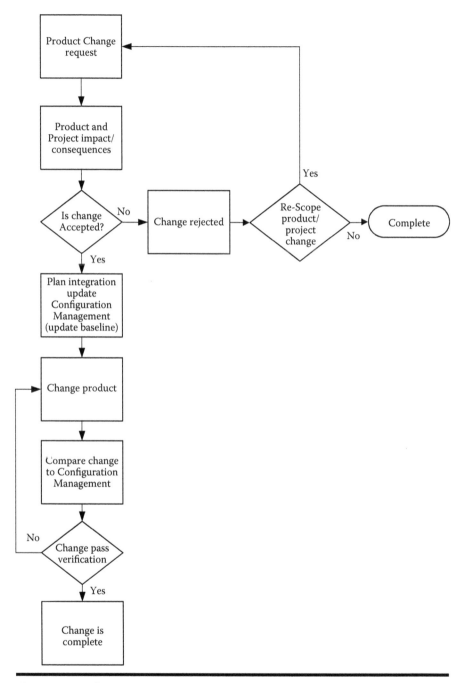

Figure 13.4   Change management.

13.3.4 Configuration Auditing

13.3.4.1 Physical Configuration Auditing

Physical configuration auditing is the action we perform to ensure that our documentation meets requirements, internal or external. In the automotive world, this kind of audit is called the *product part approval process* (PPAP), although a PPAP can also include the hardware and software. When we perform this audit, we will first verify that the deliverable documents actually exist, electronically or on paper. Once we know that the documents exist, we can spot-check the documents to verify that they meet the document specifications of the enterprise as well as those of the customer.

Physical configuration auditing will also apply to training documentation and service documentation that our customer may have requested.

13.3.4.2 Functional Configuration Auditing

Configuration auditing is the process or activity that reviews a particular delivery of the physical and functional configuration and compares both of these to the configuration expected. A physical configuration audit compares existing documents to contracted or required documents, and a functional configuration audit verifies functionality against requirements. The results are part of the release notes.

Figure 13.5 shows the five main components of any configuration management plan. These components were defined in MIL-STD-973 and have stood the test of time, showing up in the standards of other countries.

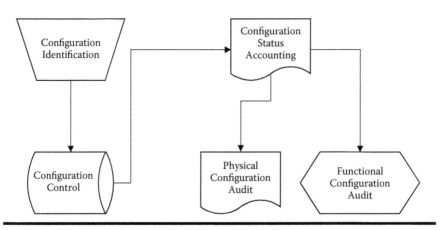

Figure 13.5   Change and configuration management.

13.4 Importance of Configuration Management

13.4.1 Release Notes

Release notes are linked to the hardware configuration by part number with the hardware content and known bugs identified. The release notes are also linked to software, again by part number, revision levels, software modules, functions/features, and known bugs discovered during the development work. The release notes are supported by the configuration management plan, with the details of the content of the specific deliveries (Figure 13.6).

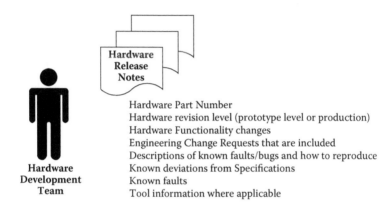

Hardware Release Notes

Hardware Part Number
Hardware revision level (prototype level or production)
Hardware Functionality changes
Engineering Change Requests that are included
Descriptions of known faults/bugs and how to reproduce
Known deviations from Specifications
Known faults
Tool information where applicable

Hardware Development Team

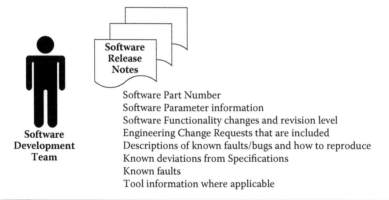

Software Release Notes

Software Part Number
Software Parameter information
Software Functionality changes and revision level
Engineering Change Requests that are included
Descriptions of known faults/bugs and how to reproduce
Known deviations from Specifications
Known faults
Tool information where applicable

Software Development Team

Figure 13.6   Release notes.

13.4.2 Identify HW/SW and System Configurations to Test Up Front

Any testing performed on a configuration must be identified explicitly with the following information:

- Hardware release, including subassemblies
- Software version
- Developmental version
- Released product version (they are not necessarily the same)

The configuration items should be linked to the test cases and test documents. Of course, the documents themselves should also come under configuration management, traceable to the versions under test. This discipline allows traceability from test cases performed to the product delivered. Having this traceability document allows for post-production feedback. For example, imagine that tests have been conducted on a particular revision of software and hardware. The assembly subsequently goes to production and the customer finds, in the field, a problem with a particular feature. It is possible to trace the test cases to the revision level and determine if the problem could have been caught during the testing, allowing for improvement of the test process and test coverage.

13.4.3 Test Plan Grows with Feature/Function Growth and Configuration

Some organizations phase the delivery of the hardware and software that comprise the product. This phasing is done to control the growth of the product, giving the maximum time with the key functions. We can use this method even if the product's full definition is not known because these early deliveries can be used to develop the downstream functional requirements. Each of these phases will have a separate configuration with details for test activities for each delivery.

13.4.4 Traceability

We know a number of reasons to desire traceability in the test and verification activities:

- Lot traceability allows us to determine if a specific lot is suspect.
- Final assembly traceability allows for a more precise recall of the product.
- Component traceability allows us to track the supplier and the supplier's specific lot.
- Software traceability allows us to track the specific software build and constituent parts thereof.

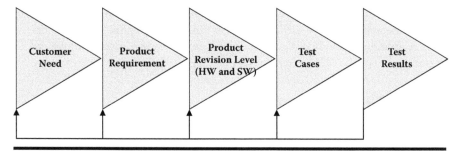

Figure 13.7   Development to test traceability.

Traceability during development and testing is as important as it is to manufacturing. We can accomplish hardware traceability with:

- Package labeling
- Radio-frequency identification tags
- Bar codes

Figure 13.7 shows how test documentation ties into the rest of the development activities. All test documentation should be traceable to specification requirements, lessons learned, or FMEA results.

13.5 Links to Testing

The importance of configuration management increases with the number of system components and the complexity of the system under test. In the event of one isolated component, no system-level interactions with other components, the links to the test suite are fewer. For products (systems) based on recurrent deliveries of a multiple of components to function, complete configuration control becomes even more critical.

13.6 Risks

13.6.1 Configuration Management Failures

When the system breaks down, we see the effect of nonfunctional components and systems. The testing activity tends to become inefficient, with increasing reports of faults that are not valid. We may also miss expected functions as well as see duplication of cost because we have to test again. We have also seen situations where a supplier shipped prototype parts with an unknown hardware version and an unknown software version—the product did not work on arrival at the customer site and the supplier later lost the business.

We say that configuration management is indispensable to delivering quality products. Failure to manage the configuration may lead to shipping the wrong product or to fixing the wrong version of software, and it generally illustrates a lack of configuration control—any of these evils can be catastrophic to a customer–supplier relationship. Good configuration management always leaves a competent audit trail.

13.6.2 Deviations

Deviations are written authorizations granted before manufacturing a product, allowing us to release the product from meeting one or more requirements. These documents only work if they represent a temporary condition we control through duration (e.g., 45 days) or by count of product (e.g., 200 pieces). They are an explicit recognition of temporary modifications outside the bounds of the change control system, although a deviation is, in a sense, part of that system. As with all forms of configuration management, deviations require an identification method such as serialization, part number control, or another marking. Inadequately accounted-for and communicated deviations can cause great harm to a project, product development, and verification because we will not have defined and documented a change and then disseminated this information throughout the enterprise.

13.6.3 Configuration Management and Outsourcing

If anything, configuration management rises in importance when contracting outsourced work, particularly with software development. In general, the more distributed the development team, the more configuration management is a priority. We can use technology to help maintain document and item integrity. The distributed team can use existing software configuration management tools as well as databases and/or Microsoft SharePoint.

Note

1. Institute of Electrical and Electronics Engineers, Inc., *Standards for Software Test Documentation* (New York: IEEE, 1999), p. 3.

Chapter 14

Software Testing

14.1 Overview

14.1.1 Qualification Testing

Qualification testing compares the requirements to the actual product in an effort to determine the degree of conformance to those requirements. The comparison is systematic and starts at the lowest building block of the software. Figure 14.1 shows the standard software engineering process that leads to testing.

14.1.1.1 Software Module

A software module is the lowest executable part of the software. A multitude of software modules are put together to meet the product requirements. Each module will perform a specific function or set of functions specified by the requirements. Information is passed between the modules via variables. Each module is a stand-alone portion of what will be the entirety of the system and can be tested as such. A developer can artificially pass these variables or input needs of the module and test the output of the module to determine if the module, in fact, performs as expected or required. This includes passing or inputting of erroneous variables where that is possible.

14.1.1.2 Software Build

The software build is the collection of software modules that will be compiled together, resulting in the software package or build. It is important to know the modules that are to be in the build and be able to trace the build to these specific modules. This is done via the configuration management system. Traceability of the build to the modules facilitates problem resolution by providing the ability to trace performance issues to a specific portion of code.

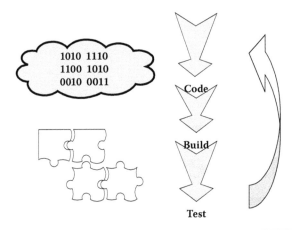

Figure 14.1   Software testing.

14.1.1.3 Software Integration (with Product)

The software constituent parts, having been tested and compiled into the software package, are now integrated into the hardware.

14.1.2 Exploratory or Stochastic Testing

As with hardware, stochastic or exploratory testing allows the test engineers to use their "gut" sense of the product's behavior to drive new test cases. The critical point during this kind of testing is to always document the preparation, execution, and results in such a way that a successful test can be added to the standard suite of test cases.

Successful exploratory testing may sometimes meet resistance from developers who prefer cut-and-dried test suites. We should never succumb to their entreaties. A constantly growing suite of test cases will ultimately put a substantial "squeeze" on the software, making it increasingly difficult for defects to escape.

The situation can get politically hot when the test engineers discover an issue that existed in previous versions of the software, a by-product of the fact that not everything is known about the product at any given time slice during and after development. We find it far better for a supplier to discover an error and report it rather than wait for the customer to discover the error and come back to the supplier with justifiable rage.

14.1.3 Combinatorial Testing

Combinatorial testing can be used with software just as easily as it can be used with hardware. In general, we will create an orthogonal array designed to provide a recipe for stimulating inputs. The primary difficulty of this approach lies with the bookkeeping required to establish criteria for correct behavior (and failure) for each treatment. We can also use pair-wise and three-wise testing in this approach—stimulating every possible pair and every possible triplet.

14.2 Software Testing—The Problem

14.2.1 *Astronomical Numbers of Test Cases to Get Complete Coverage*

Software complexity immediately yields a combinatorial explosion as the combinations grow exponentially. Some options tried over the years are as follows:

- Give up.
- Use usage statistics to drive the level of testing.
- Organically grow the test suite.
- Derive test cases formally with logic.
- Use test-driven design.
- Automate testing so we can conduct tens of thousands of test cases.
- Quasi-automatic test generation as in tree and tabular combined notation (TTCN).

14.2.2 *Development Philosophy Can Affect*

If the development team does not use a technique such as test-driven design, it is possible that we develop quality-inspector-itis and simply send their work to the test group to find the defects. We think this kind of "throw it over the wall" philosophy is dangerous. We expect professional software engineers to spend part of their day performing unit testing at the very minimum. By the time the test group receives the developmental release, the software validation should not be detecting large quantities of defects. We still recommend modeling the defect arrival rate against a Rayleigh distribution to get a feel for where we are in the continuum of software maturity.

Boris Beizer cataloged software errors some years ago. Figure 14.2 shows the allocation of defects based on Beizer's analysis.

14.2.2.1 *Test, Analyze, and Fix*

The test, analyze, and fix (TAAF) mode for development occurs when the development team has not matured enough to be able to build their software using formal methods. In short, we test until we find a failure and we manipulate the software until it works. This approach is sometimes a pragmatic method for allowing continued development but it should be obvious that it is less than optimal.

It is not enough to test and find these problems. There must be a method of arriving at the "root cause" and making the necessary corrective actions. This may

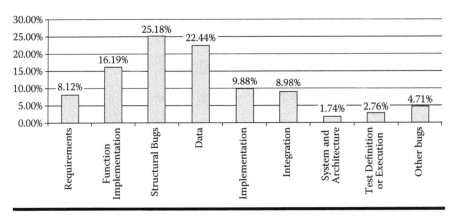

Beizer's Bug Distribution

Figure 14.2 Software defect distribution.

sound simple; however, in reality, it is not, due to the inherent complexity of the software. Finding the real root cause is one aspect of the analysis portion of the process. This is seldom as quick as most organizations make it out to be. If we do not find the real cause, the problem recurs. The goal here is to not only correct the issue but also to educate the software development team about recurrent issues; in essence, to implement a defect prevention program.

Assuming we implement TAAF (see Figure 14.3) as a formal reliability growth program, we should also implement milestone or periodic checks to assess the

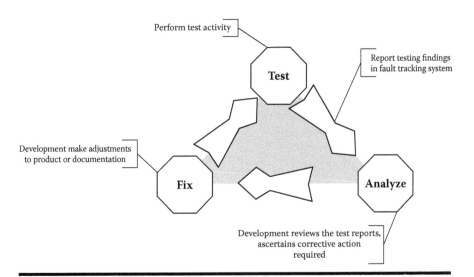

Figure 14.3 Test, analyze, and fix.

status of our program. One way to check on status is through a periodic review system:

- Test reviews:
  - We can hold a test readiness review sometime around 1 week before the start of the TAAF sequences to ensure that our preparations are in order. This review should include the following, at a minimum:
    - Results of most recent reliability program forecasts
    - Developmental status of design
    - Applicable results for tests already completed
    - Revisit open problems and failures
    - Availability of documented test procedures
    - Readiness status of test equipment and test chambers, to include functionality and availability
  - Weekly status reviews: If we are using reliability growth plots, we can update the test group on progress with these.
  - Daily status reviews: If we are using a scrum project management approach, we can increase the tempo of accomplishment.
  - Special status reviews: If we are showing insufficient progress against our reliability growth goals, then we may have to implement special actions.
- Actions upon failure recognition:
  - If we have a failure, the equipment under test should be repaired and placed back on test if we are in highly accelerated life testing (HALT) testing or a specifically designated TAAF mode.
  - Failed parts must be analyzed offline while avoiding any impact on the test schedule.
  - Offline analyses avoid stoppage of the TAAF sequence with the subsequent delays induced by analysis.
  - When we have corrective actions ready for installment, we can update the product with meaningful and controllable blocks of changes.
  - The new version of the product must then be tested sufficiently to ensure that the observed failure modes have been fixed and we have introduced no new failure modes.

On finding the purported root cause—as we implement our offline analysis—we initiate a formal corrective action, altering the software or hardware and necessitating a subsequent release, assuming that a corrective action is necessary. Because causation is difficult to prove, we may never get to a so-called root cause. In TAAF development scenarios, we may not find root cause due to

- Inadequate time
- Poor understanding of the underlying software problem
- Substandard training of the developers

14.2.2.1.1 Iterative

Finding the root cause and updating the product, whether software or hardware, is a recurrent reality during software development. In more sophisticated organizations, these subsequent releases would be delivered with release notes to the verification group. The problems found are categorized, prioritized, and the corrective actions generated to close the fault report.

14.2.2.1.2 Closed loop

When we deliver the corrective action recommendation, we must confirm the corrective action functions as expected and that it does not introduce new flaws into the product. This action closes the control loop on the specific problem. This closing action is typically performed by a verification group, who may employ actual testing of the scenario that generated the previous fault, or they may make use of simulations or calculations. If the change is severe enough, confirmation of this correction may require a larger scope to test than a single failure point. For example, imagine that a fault was reported that traced back to the operating system. Alteration or updating of the operating system has an impact on all product features. We suggest that this situation warrants further deliberation if the nature of the corrective action requires expanding the scope of the confirmation testing.

14.2.2.1.3 Reliability growth

Reliability growth occurs when we make design modifications to a product that improve the robustness of the product behavior. Two approaches to calculate the reliability growth are the following:

1. Duane model
2. Crow-AMSAA model

If we are still in the product development phase, we need to test often enough after design changes to verify whether or not reliability growth has occurred. Figure 14.4 shows a typical appearance for a reliability growth graph with its standard log-log axes. The tailing off of the cumulative failures tells us that our reliability is improving.

14.3 Test Metrics

14.3.1 Cost of Test

Costing for testing is not a trivial task. One method is to use standard cost accounting techniques to come up with a cost; another approach involves using time-driven, activity-based costing. We recommend that the organization use both, with activity-based costing as a sanity check on the conventional calculation.

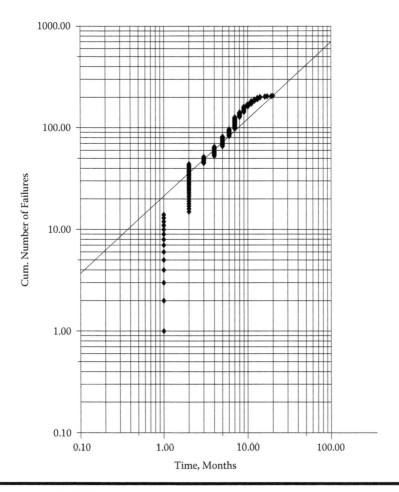

Figure 14.4 Reliability growth.

14.3.1.1 Actual Cost

The first step in a cost analysis involves checking our monthly (or weekly) expenses report, including payroll values. We also want to think about other considerations:

- Billing for generic time or actual human involvement time
- Establishing standard time for each test
- Machine-significant testing
- Human-significant testing
- Equipment degradation
- Depreciation

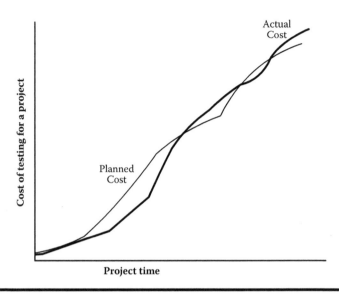

Figure 14.5   Planned cost versus actual cost testing.

- Maintenance costs:
  - Routine maintenance (e.g., acquisition of gas or salt)
  - Breakdown mechanical maintenance
  - Calibration maintenance
- Effects of staffing changes

14.3.1.2 Actual Cost Compared to Estimated or Planned Cost

If we consider our test suite as a kind of project, we can use project management tools to help us with budget, schedule, and quality issues. One method compares our actual consumption of money with what we said we were going to do. Figure 14.5 shows an example of what our metric would look like as it is plotted over time. We can use any of our project management metrics as support information for subsequent quotations to potential clients.

Figure 14.5 represents the planned cost to perform testing on a particular product compared to the actual cost of performing the test. There were two loops of testing performed on this graphic; the bulk of the preparation time is contained within the first loop. There is a second loop of additional tests; however, the preparation has largely been handled in the first loop (Ex: test fixtures developed and test cases written).

Automation of testing can offer economic benefits to the enterprise with respect to the verification and test cost. The cost to develop the fixture is the material cost plus labor to develop or to program the fixture—make it work. Increments of labor

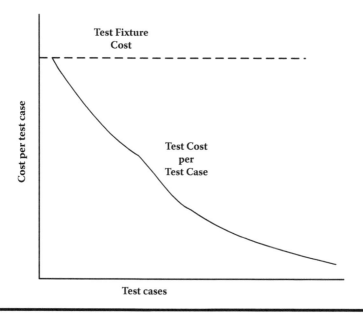

Figure 14.6   Cost per test case.

build on a base that often can be reused. If our product development model is incremental or repetitive, the previous test cases can be reused, assuming that we test any subsequent iterations or regression testing. Every time the developed test cases are reused, the unit cost for the test case decreases.

Other test cost metrics include

- Faults found per test hours
- Total number of faults to total test cost

14.3.2 Failures Found

The fundamental purpose of testing is to find any faults within the product. Finding the faults is not only the beginning of corrective action, but the frequency and type of fault are fundamental metrics, whether we are testing software or hardware. Figure 14.7 illustrates the various attributes of a fault tracking system.

14.3.2.1 Number of Open or Found Failures

A found failure is exactly that—a failure found during testing. An open failure is a failure found for which we have no response from the development organization.

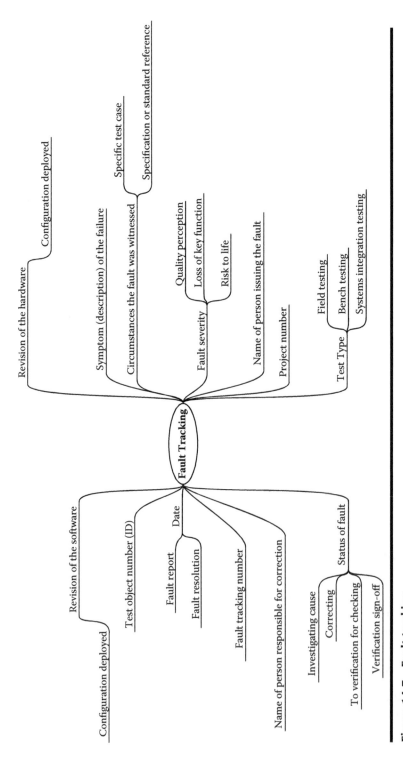

Figure 14.7 Fault tracking.

Table 14.1 Severity Table

Likelihood	Severity			
	Minimal	Minor	Severe	Extreme
Unlikely	1	1	2	2
Somewhat unlikely	1	2	3	3
Likely	2	3	3	4
Highly likely	2	3	4	4

A closed failure is a failure for which we have a response from the development team, an implemented correction, and verification that the implementation behaves as expected (i.e., to specification).

As noted elsewhere, we can also plot software failure counts versus time and use a Rayleigh model to provide some level of assurance that we are indeed prepared to release our software. However, we must proceed farther than simply plotting our defect arrival rate against the Rayleigh plot—we must also ensure that we are *increasing* our diligence by using the multi-pronged approach to forcing the product to fail.

14.3.2.2 Severity

If we are practicing automotive, pharmaceutical, food, or medical quality, we can use the results of our failure mode and effects analysis (FMEA) documents to provide values for the severity of the failures as identified in those documents. If the failures are not identified in the FMEA, we should update the FMEA with what we learn from the test samples. Additionally, we can always use the FMEA designations as a preliminary estimate of the severity of a failure.

Table 14.1 shows one way to evaluate severity and likelihood, often called *criticality*. Such tables generally bear a family resemblance. They provide a quick, qualitative method to assign a criticality value to a defect.

In any case, high severities must *always* be considered, as a single failure can lead to legal action in addition to loss of life.

14.3.3 Average Cost per Failure Found

If we find defects and we are working for a U.S. corporation, we may be able to use the estimated dollar value put into the required warranty reserve, which represents cash we could have otherwise spent on something more interesting than warranty parts. In other cases, we can look at the labor hours spent to correct a failure (cost of poor quality).

As with any area associated with quality, our cost of poor quality will include the following:

- Appraisal costs in the form of audits, inspections, and the like
- Prevention costs in the form of poka-yokes and other controls
- Failure costs:
  - Internal failures leading to scrap or rework
  - External failures producing customer dissatisfaction

The test organization can be part of both appraisal and prevention costs. Certainly, every time we test, we are consuming resources and producing an appraisal in the form of a test report. If our testing strategy also drives the development of prevention controls, particularly those in the form of mistake-proofing (poka-yoke), then we are saving money for the enterprise in the long run.

14.3.4 Test Cases Documented

Any test cases discovered during exploratory/stochastic testing should be documented and added to the formal test plan. Based on this idea, we can expect any test plan to grow during the development of a product—an indication that we are learning more about the product and its behavior. Figure 14.8 shows what a real set of test cases for hardware-in-the-loop software testing can look like. The acronyms are intentionally cryptic to protect the customer's product.

14.3.5 Test Cases Executed

We would actually look at test cases executed versus test cases planned. In general, a failure to execute a complete suite of tests should be documented as part of the test report; sometimes, we have a legitimate business reason for not executing all the test cases. For example, if we are in a test-analyze-and-fix scenario where we are providing a closed-loop correction system, we may choose to test only the parts of the product that have been modified or added to. This choice is particularly apt when dealing with software; however, we can also apply this reliability growth approach to hardware and manufacturing processes. Clearly, we would run a full validation of the product before any significant release to a customer.

The use of "test cases executed" can become a meaningless metric if we are stuffing our test documents with redundant compliance tests. When compliance testing, we should test up to the point where we have proven that we meet the specification and then move on to the next text. With exploratory, combinatorial, and severe environment testing, we can push the part as far as we need to go in order to generate failures and, eventually, product destruction.

ID	INITIAL CONDITION	ACTION	EXPECTED RESULTS	P/F
Power on Reset				
1	• Battery off • Ignition off • No service tool attached to J1939 datalink • Configuration checksum is set to a know value	• Turn battery on • Send the following messages during power on reset DM14 18d917f9x 07 05 04 0a 00 00 8d 3a • Wait 1 second DM16 18d717f9x 07 00 00 80 FF A5 A5 A5	• EGC enters into bootloader session after 5 seconds • After 1 second the cluster should to send the following DM15 DM15 07 08 00 00 FF FF FF	P
2	• Continue from above	• Send the next message to force a jumping from bootloader to operational program. DM14 18FF30F9x 19 00 40 00 00 FF FF • Wait 1 second Turn battery and ignition off	• The cluster should to be off.	P
3	• Battery on • • Ignition on • Communication data link J1939 sending.	Send the appropriate messages to avoid faults • Drive all the gauges to different positions	• All gauges shall respond correctly to the message sent	P

Figure 14.8 Test case documentation.

Stand by mode

#			P
4	• Continue from above	• Turn ignition off • Send the following messages during power on reset DM14 18d917f9x 07 05 04 0a 00 00 8d 3a • Wait 1 second DM16 18d717f9x 07 00 00 80 FF A5 A5 A5	• EGC enters into bootloader session after 1 second • After 1 second the cluster should to send the following DM15 DM15 07 08 00 00 00 FF FF FF	P
5	• Continue from above	• Send the next message to force a jumping from bootloader to operational program. DM14 18FF30F9x 19 00 40 00 00 FF FF • Wait 1 second Turn battery and ignition off	• The cluster should to be off	P
6	• Battery on • Ignition on • Communication data link J1939 sending.	• Send the appropriate messages to avoid faults • Drive all the gauges to different positions	• All gauges shall respond correctly to the message sent.	P

Memory Write X0A

7	• Continue from above	• Turn ignition off • Send the following messages during power on reset DM14 18d917f9x 07 05 04 0a 00 00 8d 3a • Wait 1 second DM16 18d717f9x 07 00 00 80 FF A5 A5 A5	• EGC enters into bootloader session after 1 second • After 1 second, the cluster should to send the following DM15 DM15 07 08 00 00 00 FF FF FF	P

Figure 14.8   (Continued).

14.3.6 Test Coverage (Test Cases to Requirements)

We can use a traceability matrix to help with test coverage. If we are an electronics manufacturer, we can also assess the number of nodes on the printed circuit board and verify how many we are testing, usually with a device like an in-circuit tester (ICT).

We may run into a situation with software development where it becomes difficult to estimate the test coverage because the flow through the code is less than obvious. With embedded software development, we find relatively few automated tools for challenging the product. Consequently, we do not usually have white-box-derived test coverage values automatically generated by the software. Hence, we can see why we use the manually maintained test traceability matrix as a tool for ensuring test coverage to the specification.

Of course, the other portions of our test strategy (combinatorial, stochastic, severe) will generally *not* be directly traceable to a specification (otherwise, they would in the compliance testing phase); hence, we cannot track these items directly as part of our test coverage. Saying that we have 180% test coverage is a meaningless expression.

14.3.6.1 Duration to Close (Average Time to Close)

This measurement assesses how long it takes, on average, to correct the error; that is, essentially it is the time to deliver the fix. This metric can include the reverification time. We have found that we frequently receive prompt responses to test incident reports but the verification portion of the incident report can often take months, particularly if a tool change or a printed circuit board change must occur for the reverification to happen.

14.4 Software Boundary Testing

Boundary testing occurs when we test the extreme values of the input domain:

- Maxima
- Minima
- Barely within the boundary
- Barely outside the boundary
- A variety of nominal values
- Deliberate error values

With embedded software development, we want to deliberately introduce errors on the data bus (among other nasty intrusions!). A data bus such as a controller area network (CAN) can be surprisingly intractable with regard to deliberate error introduction: Most of the tools do not permit erroneous messages, and the CAN

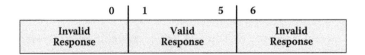

0	1	5	6
Invalid Response	Valid Response		Invalid Response

Figure 14.9   Software boundary testing.

mailbox system works against errors. Hence, the test team will often have to develop its own embedded tool, hardware, and software to introduce noise-affected messages to the data bus.

Boundary testing is simple to understand but often difficult to implement. Regardless, we see no excuse for not pushing the product to its limits, especially at the boundaries.

14.4.1 Maximum and Minimum Values

14.4.1.1 Lookup Table Ranges

One way to introduce maxima/minima testing will involve a lookup table. If we have defined every input and output for our system, we need only catalog these in a machine-readable table and let our test software execute the appropriate values.

14.4.1.2 Analog and Digital Input Ranges

The concept of electrical analog value testing is fairly straightforward. We know the maximum and minimum expected signals, and we know that nominal values lie between these extrema. We can use a combinatorial recipe or some other approach to stimulate these values.

Digital value testing can be interesting if we are able to stimulate the product with values that lie somewhere between electronic zero and electronic one—an indeterminate region. We are then looking for the reaction of the system to an indeterminate truth value. A digital failure explains the reason we use watchdog timers to reset the system. Indeterminacy can lead to an apparent system "hang."

14.4.1.3 Function Time-Out

With some embedded systems, we will deliberately introduce a timer with specific functions—they are *supposed* to time out. Obviously, our testing should put us in a situation where we can challenge the time-out mechanism. One way to do this is to attack this particular function with an unexpected sequence of interrupts. If the clocking for the time-out in the function is not deterministic, we can expect a software failure. Basically, when we use interrupts this way, we are forcing a high level of context-switching with the attendant time consumption required to save registers and stack frames (if we are doing those) and the subsequent return to the original state, only to be interrupted again.

14.4.1.4 Watch-Dog Performance

A "watch dog" is used when we have an embedded software product. External watch-dog timers must be reset periodically to keep them from resetting the entire system. We often call internal watch-dog timers "watch puppies" because they are not usually as robust as the external devices, although they function the same way. Basically, if our software does not reset the watch-dog timer in time, we assume that the software has become too incoherent to function properly. Once the device software is allegedly stable, the watch-dog timer is often turned off by mutual agreement between the developers and the customer.

14.4.1.5 Transaction Speed

We can trouble transaction processing with interrupts in much the same way as we tortured the function time-out capability. Interrupts always force software to become asynchronous, driving down predictability and causing unexpected problems. We have only been able to achieve synchronous software a few times, after which the embedded world became a beautiful place—all activities were now predictable. In most realistic cases, our testing will drive to bog down the processor enough that it begins to have trouble performing routine processing, which, in turn, becomes noticeable to the end user.

14.5 Static Code Analysis

Static code analyses are the things we do to determine the capability of the software without engaging the software functionality.

14.5.1 Test without Testing

We use static code analyzers whenever possible because they are generally quick and they provide a first-pass listing of real and potential problems in the software. An example of this kind of tool is the well-known *lint* processor developed originally on UNIX systems. It reads through the textual code file and checks for common programmer mistakes, after which it generates a report. The tool has numerous flags that allow the test team or developers to customize the rigor of the analysis. If the developers are not trying to convert a language like C to a heavily typed language like Ada, they will modify the tool appropriately.

14.5.1.1 Control Flow Analysis

Control flow analysis reviews the flow of the various functions within the software. We review how the software modules are put together to produce the desired feature content and performance. Most control flow analyses involve mapping the flow of the logic of the program. The McCabe cyclomatic complexity calculation is a variation on control flow analysis.

14.5.1.2 Data Flow Analysis

Data flow analysis requires a review of the source code with the objective of understanding the values possible at any point in the code execution. We review the contents of variables throughout the function, calculating the values at each execution point that can modify the data, predicting the values that will be produced during the execution, and comparing those values with what are believed to be valid values. The data flow diagram is one tool we can use to model the flow of data in our code.

14.5.1.3 Conformance to Identified Coding Standards

Code beautifiers can help modify the source code to meet identified coding standards. Some static checkers can also perform this function. Over the years, these tools have had names such as "indent" and "prettyprinter." The beautifiers not only help enforce standards; they also usually make the code consistent in format and therefore much easier to read.

14.5.1.4 Checking for Known Syntactical or Logical Errors

Syntactical errors are often detected by the compiler. This is also true for some of the logical errors but to a lesser extent. We can also use a static checker to look for syntactical errors that are more complicated.

14.6 Dynamic Code Analysis

Dynamic code analyses are the activities we perform on the software once it is compiled and we exercise or engage the software.

14.6.1 Logic Analyzers Can Perform This Function

We usually use a logic analyzer when we have an apparent timing problem. Logic analyzers are convenient tools for analyzing the timing of multiple channels simultaneously. More sophisticated logic analyzers can also perform some level of code disassembly, giving the tester a window into the steps as the code precedes around the time of interest. The most capable logic analyzers can

- Perform protocol decoding
- Convert captured data into timing diagrams
- Disassemble

- Pair assembly code with source-level software (!)
- Trace machine states

Modern logic analyzers use probes much as oscilloscopes do. The test engineer will set up a sequence of *triggers* that will start the logic analyzer recording a sequence of channels. The best logic analyzers can save and export these values for subsequent analysis using personal computer software. Logic analyzers have the ability to reveal hardware defects that would not be found in emulation or simulation. Part of the reason for this benefit occurs because, with the logical analyzer, we are probing an actual product rather than a virtual product (simulation) or a hybrid product (emulation).

14.6.2 In-Circuit Emulators Also Do This

Hardware in-circuit emulators (ICEs) allow the software engineer to load programs into the embedded system, execute them, and then step through them slowly, viewing and altering the data used by the embedded system. In a sense, they behave very much like a debugger but with the addition of special hardware support for the embedded microcontroller. Most modern ICEs will have a "pod" into which we place the microcontroller. The pod allows us privileged access to the target processor while giving the software engineer the control of a debugging tool. The ICE can help give the software engineer some clues as to the source of the observed failure, a benefit that can be difficult even with logic analyzers. Even getting to the point where the developer knows which *section* of code failed is a huge leap forward in productivity.

As with software debuggers, the ICE will generally provide the developer with breakpoints, memory triggers, code watches, and input/output control. More advanced versions may have watch criteria defined by the software engineer. With some modern microcontrollers, it may be possible for the ICE to use special features of the microprocessor to lessen the cost of the emulator hardware (e.g., Microchip Technology Inc. has done this with its PIC® product line).

14.6.3 Debuggers Have Dynamic Behavior

We rarely use debuggers with embedded code—we nearly always try to use an in-circuit emulator. We have gotten this far without establishing why debuggers and emulators are significant to testing. They can be used to *step* through the code, which is a form of unit testing with special observational capabilities.

14.6.4 The Capability Allows for in Situ Analysis of Code Behavior

We use logic analyzers, oscilloscopes (when we have to), ICEs, and debuggers to give ourselves a chance at seeing the code behavior line-by-line, action-by-action.

Without these tools, we can spend significant amounts of time guessing at the cause of specific behaviors. The tools bring us closer to the code but with some limitations:

- Emulators with their own power supplies may not react the same way as a chip.
- Logic analyzers must have high-speed sampling.
- Logic analyzers generally need expensive supporting packages to be able to represent code as code.
- Debuggers are limited when using embedded software.
- Oscilloscopes require significant skill and usually have insufficient channels.

Note

1. Institute of Electrical and Electronics Engineers, Inc., *Standards for Software Test Documentation* (New York: IEEE, 1999), p. 3.

Chapter 15

Systems Testing

Systems testing is also known as *integration testing*. This sort of testing can be extensive, covering many newly developed modules, both from the software and hardware perspectives. As the name implies, at this point, we are testing an entire system. We can expect responses from both main factors and significant interactions.

15.1 End-to-End Testing

One type of systems testing is known as *end-to-end testing* or *E2E testing*. In this case, the test will spread across a number of systems. For example, a telemetry system on a vehicle interfaces with the vehicle but also transmits vehicle information off the vehicle to a central station. This information can be viewed over the Internet to a specific customer location. In this instance, the E2E testing will consist of the vehicle integration, the transmitted information of the results to the display, and calculations on an Internet location requiring a number of tools.

Simon Stewart's article, "9 Tips to Survive E2E Testing," provides good guidelines for handling this form of testing.[1]

1. Get involved early.
2. Do not start your testing until the application's dependencies are ready.
3. Run your tests at different scales.
4. Automate as much as makes sense, but no more.
5. Automate as much as makes sense, but no more (Yes, Tip 5 is the same as Tip 4).
6. Find another way to test the code.
7. Delete tests that are no longer relevant.
8. Do exploratory testing.
9. Rely on your automated tests

15.2 Big-Bang Testing

With this method of systems integration testing, all of the subassemblies or components are assembled just as the end system is envisioned. The testing starts with the system fully constructed as if it were a sellable product. This approach does not account for the subassembly tests to determine conformance to specifications. In this method, the root cause of any system performance issue must be sorted out from the system interactions, which can be quite complicated. The original goal of saving time and money is lost. The benefit—if it could be called that—is that as soon as the subassemblies are available for testing, they are tested.

Usage model testing is similar to the big-bang approach. We will assault the product, software or hardware, with a simulation of the expected usage of the part. In some cases, we may use a statistical approach, where portions of the design that are less likely to be used are also tested less. The risk in this approach is obvious, not to mention that we have already discussed how difficult it is to produce a truly representative usage environment.

15.3 Top-Down Testing

Top-down is not the same as big-bang testing. In top-down testing, the system constituent modules are tested at the high level of functionality progressing downward. For example, we would make sure that the serial bus communicates valid messages and then we would work down the hierarchy, verifying the bit timing and bus access times.

Functions that do not exist as yet can use substitutes we call *stubs*. A stub routine might report that it has been called, and that is all it does until the correct function is developed. We most likely will, in fact, start at the very top with a main routine and the rest of the code in stubs. Once the main routine functions correctly, we begin to replace stubs with real functions. Please note that this modality implies that we have a fully developed architecture and we know where we are going with the design.

Using the top-down approach allows us to do the following:

- Provide some level of demonstrated capability to ourselves, our management, and our customers.
- Achieve early success (the illusion of a real application).
- Implement our architecture as an architecture rather than as a loose federation of functions.

The top-down approach has some drawbacks, however, and some of these are the following:

- The illusion of success (that is, our test condition is a kind of false world)
- Difficulties verifying the code as the stubs are replaced by real functions
- No or diminished interaction effects (side effects) at the beginning

15.4 Bottom-Up Testing

Bottom-up testing starts with the smallest elements of the design, which, again, means that we must have a complete understanding of the architecture of the final product. If we consider a system comprising a number of controllers, we would focus on one of those controllers. The testing would start at the smallest constituent parts of a particular functionality within that controller—we sometimes call these terminal functions. For example, we would test the timing of bits from out of the serial data bus of a specific device under test.

Because we are testing terminal functions at the start, we have some drawbacks with this modality:

- No top-level architecture exists at the start of testing.
- Instead of stubs, we must have driver functions that call the terminal functions, again providing a false world.
- No true product until the end.

On the other hand, bottom-up testing/development can provide the following benefits:

- We can verify that the terminal functions work, which is essential in embedded code if they are hardware drivers.
- Tests are easy to create.
- With embedded code, we start with the hardware and move up to the main routine.

15.5 Managing Multiple Test Hardware and Software Deliveries

Concurrency is always a trying experience. We can simplify the approach somewhat by doing the following:

- Establish the points where the deliveries must be simultaneous.
- Do this step for more than one point if more than one set of simultaneous deliveries is applicable
- *Back-schedule* from each coordination point.
- Make sure that we have some slack in the plan or it will never happen as expected.

Back-scheduling starts with the realization that we are most interested in the delivery date, not in the start date for these multiple threads of activity. Neophyte project managers often make the mistake of building their plans/schedules/timelines from the start date rather than from the other direction. The "secret" here is the back-scheduling, much like that of a modern manufacturing resource planning (MRP) system, plus the added safety leadtime of additional slack built into the plan.

We want to ensure that our plan has enough slack time built into it to make it much more likely that we can make coordinated and timely deliveries. An experienced project manager will monitor total slack carefully as a means of checking on the real status of the project. If we also have calculated separate total slack times for the threads, we can also monitor and manage those. Once the slack is consumed, our project enters a critical path, which is generally a catastrophe to delivery times. Any delay on the critical path will delay the project.

Figure 15.1 shows a very simple representation of what the first step in back scheduling would look like. When using software such as Microsoft Project, we should remember to set up the software to facilitate the back-scheduling approach.

15.6 Test Configuration Management

Just as with software development, all of our testing, whether hardware or software, should fall under some kind of configuration management. This requirement means we have the standard four components (Figure 15.2):

1. Configuration identification
2. Configuration control
3. Configuration status accounting
4. Configuration auditing:
   - Physical configuration audits
   - Functional configuration audits

We described general testing configuration management with Chapter 13; however, we usually need to take some extra steps when working with software testing. For example, if we have multiple branches of a specific version of software, we must manage these different versions and verify that we are testing the correct version when we execute our test plan. When possible, we advise having the software send us a message telling us which version it is—this approach becomes a kind of prime verification for the software under test.

15.7 System Variations and Test Demands

We mentioned that we must control for variation in the software, but we also must control for different hardware versions. Using a pairwise, combinatorial arrangement we achieve what we see in Table 15.1. This table shows a situation where we match up four versions of software with nine versions of hardware and execute each pair at least once.

ID	Task Name	Start	Finish	Duration	Mar 2010														
					17	18	19	20	21	22	23	24	25	26	27	28	29	30	
1	Hardware development	3/25/2010	3/30/2010	3d 4h															
2	SW testing	3/23/2010	3/30/2010	6d															
3	Hardware testing	3/29/2010	3/30/2010	2d															
4	Software development	3/19/2010	3/30/2010	7d 4h															
5	Documentation	3/30/2010	3/30/2010	1d															

Figure 15.1   Multiple deliveries.

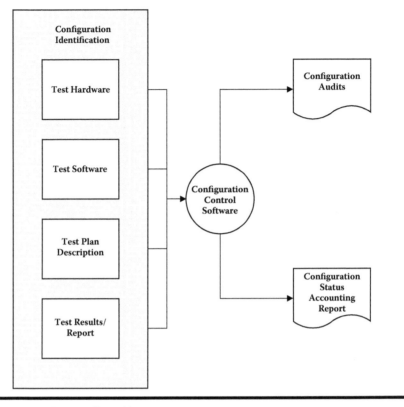

Figure 15.2　Test configuration management.

15.8 Functional Testing

In general, when we see a term like *functional testing*, we are performing one of the following options:

- A black-box test
- An end-of-line production final tester (as opposed to in-circuit testing in production)
- A functional configuration audit

We will normally be testing at least a subsystem and, often, the complete system. During functional testing, we will verify compliance to requirements at a minimum.

Table 15.1   System Variation and Testing

Case	Software_version	Hardware_version
1	2.3a	8280
2	2.3b	8280
3	2.3c	8280
4	2.3d	8280
5	2.3a	8380
6	2.3b	8380
7	2.3c	8380
8	2.3d	8380
9	2.3a	8480
10	2.3b	8480
11	2.3c	8480
12	2.3d	8480
13	2.3a	8580
14	2.3b	8580
15	2.3c	8580
16	2.3d	8580
17	2.3a	8680
18	2.3b	8680
19	2.3c	8680
20	2.3d	0680
21	2.3a	8780
22	2.3b	8780
23	2.3c	8780
24	2.3d	8780
25	2.3a	8880
26	2.3b	8880
27	2.3c	8880
28	2.3d	8880
29	2.3a	8980
30	2.3b	8980
31	2.3c	8980
32	2.3d	8980

15.9 System Response to Multiple System Failures

In the event that multiple subsystem failures occur simultaneously, each failed part that would independently deter satisfactory product performance becomes a relevant failure except when the developers and the test team agree that the failure of one part was entirely responsible for the failure of any other parts (cascading failure). In this case, each associated dependent part failure is *not* counted as a relevant failure. At least one product-relevant failure must be counted when we claim a dependent failure.

On the other hand, failures due to defective subsystems are also relevant failures. If several component parts of the same classification fail during the test, each one is considered a separate relevant failure unless it can be shown that one failure caused one or more of the others as in the previous paragraph.

15.10 Ranges of System Performance

By now, we know that we must test the entire system across the gamut of nominal behaviors. However, we also suggest that the system-level design failure mode and effects analysis (DFMEA) be used to drive a further set of tests designed to verify that we have eliminated any critical issues identified in the DFMEA. If we have time and resources, we would also push the system testing to the extreme level. That way, were an untoward event to occur later in the life of the product, we would have enough system characterization to know if the product had been abused.

Note

1. Stewart, Simon, "9 Tips to Survive E2E Testing," *Software Test and Performance Magazine*, 7(4), 8–12, April 2010.

Chapter 16

Simulation and Emulation

There are other tools that are used in the quantification of design performance—simulation and emulation. When these tools are coupled with testing, we can go a long way to proving and improving the subsequent design. Through their use, design and functionality capabilities are exercised well before the entire product is available. In this chapter, the discussion centers on their use and how they complement and supplement one another in securing the design quality.

16.1 Simulation

Both simulation and testing have specific goals. For simulation, we want to facilitate requirements derivation, uncover unknown design interactions and details, and reduce development costs by requiring fewer actual parts. Much of this activity facilitates testing by quantifying the requirements, developing test cases, and generally making testing more productive (see Figure 16.1). For testing, we want to achieve defect containment, reduced product warranty cost, and some level of statistical indication of design readiness.

16.1.1 Simulation Typically Refers to a Model of a Process or Function

Simulation generally refers to a model of a process or function; for example, we can simulate the general behavior of a manufacturing process, a motor vehicle, or any other object for which we have knowledge about inputs, outputs, and behavior. Wedding simulation with automated testing allows test organizations to achieve benefits such as increases in testing speed (throughput), increases in test coverage for both hardware and software, and the ability to test before hardware becomes

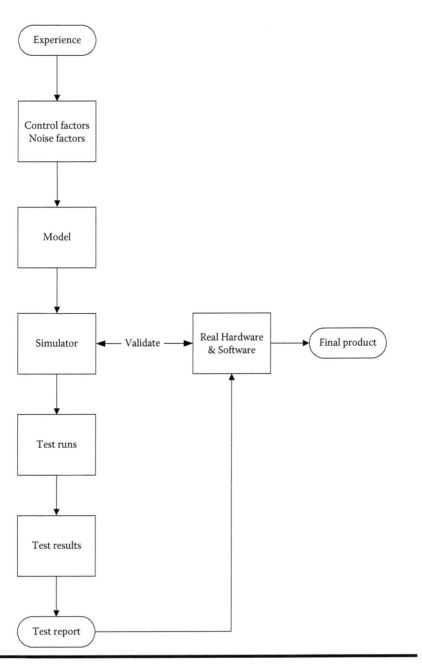

Figure 16.1 Simulation.

available. In this section we describe each type of approach in turn and then how they can work together synergistically.

Both simulation and testing have specific goals. For simulation, we want to facilitate requirements generation, uncover unknown design interactions and details, and reduce development cost by having fewer actual parts. Much of this activity facilitates testing in quantifying the requirements, making testing more productive (see Figure 16.1). For testing, we want to achieve defect containment, reduced product warranty cost, and some level of statistical indication of design readiness.

16.1.1.1 Example: We Can Simulate the General Behavior of a Manufacturing Process

Figure 16.2 provides a high-level approach for creating a full-scale simulation of a manufacturing process. In most cases, we are likely to use a discrete event simulator to create our simulation. Commercial products such as Arena® are designed to do this or we can write our own version using a programming library designed to support simulation. For example, SimPy is designed to be used with the Python language to create discrete event simulations.

By creating a simulation of the manufacturing process and validating it against the real process, we are now in a position to assess the effects of control factors and noise on the behavior of the line—without ever having to interfere with the real line! Even experimental techniques such as evolutionary operations or sequential simplex require adjustments to bring the model into alignment with reality. With simulation, we test our changes in a virtual factory of our own creation.

16.1.2 System

When we use a system-level simulator, we generally model the system more crudely than when we are looking at component-level simulation. Sometimes an executable block diagram (e.g., BlockSim® by ReliaSoft) may be sufficient to provide enough simulation capability to deliver what we need to know.

When one of us worked in the defense contracting business at a test range, we always wrote the simulator before writing the main code for the application. For example, when developing a simulator for main battle tanks, we used the following sequence:

1. Create a batch simulator that uses canned code to drive the tank around the track.
2. Create a high-level simulator that does the same thing but can operate with user input.
3. Create a more detailed simulator that has all the communications and other functions we would expect on the actual vehicle.

In general, if we are developing both an application and a simulation for that application, we do *not* want the same developer doing both sets of code. If we use the

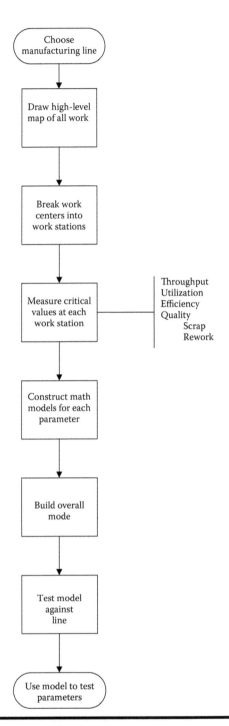

Figure 16.2   Manufacturing simulation.

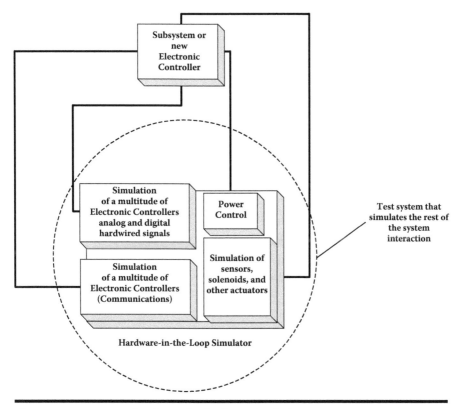

Figure 16.3 Systems hardware-in-the-loop simulation.

same developer, we can be deluded that our system is functioning correctly when, in fact, it only functions in its own false world.

System simulation not only facilitates requirements generation/derivation and definition of the end product, but it can also be used to replicate the entire system and serve as a test function. We use hardware-in-the-loop simulation to fill this need. The subsystems are either attached to the hardware simulator or can be coded into the simulation. Heavy-duty vehicle manufacturers can have large variation in the vehicle configurations, and a tool such as this allows for verification of these system variations without the need to build each of the vehicle configurations. Many have developed systems that can either simulate the rest of the vehicle systems or that allow for the addition of subsystems to the test fixture in order to simulate the system. In Figure 16.3, the area within the dotted-line circle is within the test fixture and replicates the various electronic controllers such as the engine control.

While this type of system makes it possible for systems testing well in advance of the total system or vehicle being available, there are limitations. This is just a simulation of a vehicle, and some correlation between this system rig and the actual vehicle needs to be established to provide some confidence in the tool actually replicating the real world.

16.1.2.1 Electrical

The best known of the electrical simulators for small circuits is Simulation Program with Integrated Circuit Emphasis (SPICE), which is designed for analog electrical circuits. More recent versions of this family of software applications can perform optimization as well as modeling. Given the ubiquitous availability of models for specific electrical and electronic suppliers, it is puzzling why this level of simulation is not more common in most new product development. Simulations with the PC Simulation Program with Integrated Circuit Emphasis (PSPICE) tool (examples of providers include MicroSim, OrCAD and WinSpice) will accommodate digital gates such as

1. Digital gates such as AND, NAND, OR, NOR
2. Multiplexors
3. Decoders
4. Flip-flops
5. Latches

16.1.2.2 Functional

For many functional simulations, we will most likely choose a discrete event package for a variety of reasons:

- Relatively easy to use
- Can be executed at quicker-than-real-time speeds
- Many packages exist

On some occasions, we may choose to use a real-time package because we are also going to use the simulator as a training tool, much like the airlines use simulators to train pilots. The simulators usually present a high level of verisimilitude so the user reacts appropriately after the willing suspension of disbelief.

Simulation of particular functions can be performed via the hardware-in-the-loop simulator. These simulations allow concepts to be tested early in the project and within the confines of a specific system. The same tool used to simulate the system also provides a verification tool to integrate early prototype parts for testing. It can be used to simulate the functions within a system. In the example below, all electronic control units are virtual or simulated; however, there is no need to restrict the test this way. The simulation could be a feature distribution with existing hardware or a system that is only partially simulated.

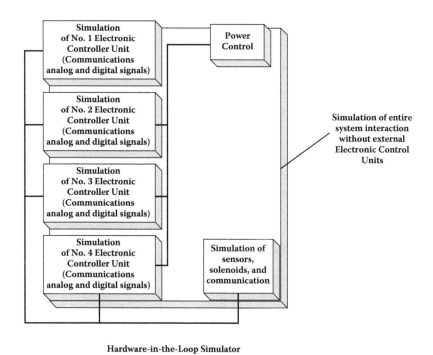

Hardware-in-the-Loop Simulator

Figure 16.4   Hardware-in-the-loop simulation at concept.

16.1.3 Component

16.1.3.1 Electrical

The tool allows the design engineer to put the schematic together, just like any other circuit layout program. For years, electronic component suppliers have provided convenient models for the components they carry. Because these models can involve a substantial development effort by the customer, the extra effort on the part of the supplier can be an order-winning part of the supplier's marketing edge. Maxim provides models for analog switches, comparators, and operational amplifiers. SPICE tools allow for simulation of the product design early in the development process, even as the schematic for the product is being developed.

16.1.3.2 Functional

We have witnessed the use of an Excel spreadsheet that was used to model the human–machine interface of an instrument cluster. The worksheets were linked using keyboard presses to demonstrate how the product would work. Functional simulations are typically used to provide a tool for gathering requirements. When this technique is used appropriately, the customer can provide some constructive input on the product.

16.1.4 Environmental

The aforementioned tools provide the ability to simulate the circuits over a range of temperatures, allowing sampling at various nodes or key points in the schematic. In this way, we learn enough about the product that we can propose design response changes due to ambient temperature variations.

Environmental simulation does not stop at variation in the temperatures the product can expect to experience; it also allows for modeling of electrical transients introduced at various points in the design. Furthermore, modeling the derating responses of various components to increased temperatures allows an evaluation of the proposed design before the product is available. It is possible for an engineer to calculate the performance of the individual components and the overall system over a thermal range. This approach is purely theoretical, and it can take a great amount of time with plenty of opportunities for errors.

16.1.4.1 Electrical

Radiated and conducted transients (bulk current injection) the product can see can also be modeled. It is possible to inject these signals into various parts of the product and observe the impact. The areas of concern are typically any power supply sections, as well as inputs and outputs of the product.

16.2 Simulation Levels

16.2.1 Virtual Simulators

16.2.1.1 Partial Hardware

Our hardware-in-the-loop system is an example of a partial hardware simulation. In this case, some of the system is simulated. However, there are possibilities to add hardware to the system at any time, including any device or subsystem under test. Hence, the partial hardware approach allows for a quick start with the simulator—we do not have to write the entire simulator because we already have some of the pieces.

16.2.2 Live Simulators

Running a live simulation is akin to running the actual product in the field or on the bench—we are in real time and events occur with roughly the same duration we would expect to see in the real world. We lose the benefit of speed but gain the

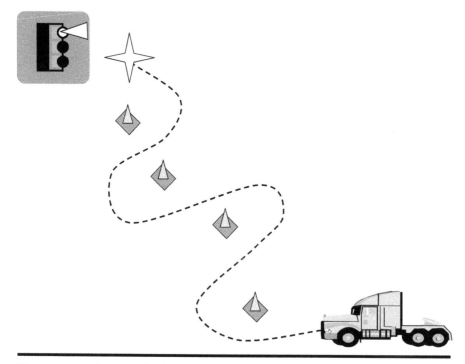

Figure 16.5   Live contrived exercise.

benefit of greater similitude with our real products. Most of the aircraft and military vehicle simulators we have used have been live simulators. Because these simulators require human intervention as part of the simulation, it does not make sense to run them faster than the real-time rate.

16.2.2.1 Live Contrived Exercises

Live exercises are used extensively in military applications. A well-known version of a live exercise is live fire testing, where actual artillery or rifle rounds are used. This method is also used in automotive and heavy machinery. When we see the car advertisements where the vehicle is performing slalom maneuvers in heavy traffic, this is an example of a contrived simulation. Figure 16.5 shows a vehicle moving through an obstacle course and stopping at a red light.

16.3 Simulation Activities

It is possible to learn much about the product early in the development process. However, to take advantage of this possibility, a few things must happen. Figure 16.6 shows the sequence of activities required. Key variables must be identified and models built. Building the models is only part of the equation. A simulation is only as good as

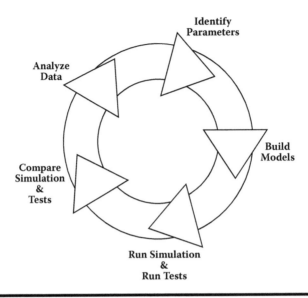

Figure 16.6   Simulation process.

the real world it represents. We must take the developed model and validate that all is well—without this self-check, the simulation would be analogous to developing a product and launching with no testing. This process is an iterative comparison of the simulation results to real-world responses. As these simulations and the real world converge, the confidence in the predictive powers of the system model increases.

16.3.1 Identify Goals

As with any task, we work to identify the goals for our simulation development. We have to consider the *type* of simulation we propose to use and then set the rest of our goals based on that or, conversely, set our general goals and pick the type of simulator that meets those goals well. We need to consider the following:

- Where in the hierarchy we are simulating (system, subsystem, part, component)
- The time basis for our simulation (real-time, high-speed, slow motion, etc.)
- Necessary hardware to accomplish the build of the simulator (level of automation)
- The software we need to build the simulator

16.3.2 Prepare for Simulation

Preparing for simulation consists of many of the same activities used to understand the product requirements. If the simulation is to be of any value, it must have some basis in the expected reality of the product application.

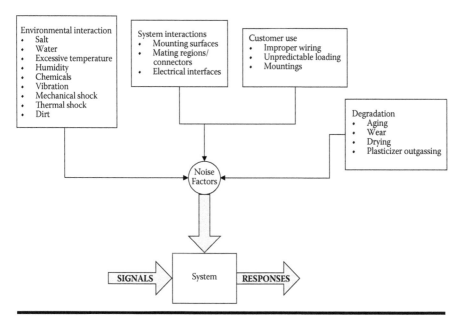

Figure 16.7   P-diagram example.

16.3.2.1 Identify Parameters

We recommend the use of an artifact such as a P-diagram to indicate signals, control factors, noise factors, and outputs for the simulation. This diagram can provide the first pass assessment of the parameters we will need to simulate. For larger systems, we need to build a functional breakdown diagram and assign a P-diagram to each function. Figure 16.7 shows a generic version of a P-diagram. These diagrams are often used to help define factors for designed experiments, and we think they can function equally as well defining parameters for a simulation.

16.3.2.2 Model Parameters

Once we choose our parameters, we need to model them. To model these parameters, we need to assess how they function within the system. We might use the following options:

- ■ Continuous state parameters:
  - – Integrators (accumulators)
  - – Transfer functions
  - – Zero poles
  - – State spaces
- ■ Discrete state parameters:
  - – Step functions
  - – Point functions

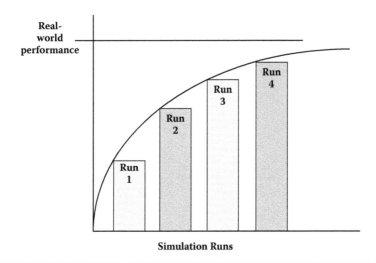

Figure 16.8   Run simulation.

- Variable-step functions
- Any gains
- Switches
- Comparators
- Algebraic functions (adders, subtractors, multipliers, dividers)

If we are using an actor-based simulator (e.g., NetLogo), we must specify some behavioral rules that the actors will use as constraints to operation. In some cases, an actor-based simulator becomes more than a simulator: ant colony–type simulation can be used as an optimizing technique.

16.3.3 Run Simulation

Usually, we will execute multiple runs of a simulation to establish a baseline of expected behavior from the simulator. This approach also provides enough information to determine if the simulator provides a reasonable representation of the real system. As we change parameter values, perhaps according to a designed experiment plan, we can perform as many replications of the experiment as needed. Figure 16.8 shows how we can use multiple runs of our simulator to become asymptotic to the real situation.

16.3.4 Perform Test

If we have programmed the simulator to allow for some level of random behavior, we can use our responses (test results) to calculate confidence intervals and maximum likelihood values. We know whether we will venture into a probabilistic approach because we did our goal setting before starting our program. If we are using accelerated simulations, we can set our runs in a loop and execute tens, hundreds, or thousands of simulations an hour—depending on the complexity of the simulation. Some tools, like Telelogic Tau (an implementation of the various ITU standards such as the System Definition Language and currently owned and maintained by IBM), can execute millions of runs in an hour, effectively assaulting the design with replicable or random or specific tests.

16.3.5 Gather Data

If we have designed our simulator well, the simulator itself should be collecting data. We need to ensure that we have sufficient storage for our results if we are performing hundreds of thousands or millions of runs with the simulator. The ability to generate vast amounts of performance data is one of the desirable capabilities of a simulator.

16.3.6 Analyze Data

We gather data in order to analyze data. We will recover the data from storage and perform an analysis of the results just as if we had tested the real product. If we have truly massive data sets, we can perform statistical analysis using a statistical programming language/application such as R. R provides nearly 2,000 packages to assist the user with analysis, and we can easily build our own custom scripts as well. If a spreadsheet is inadequate for the task, we might consider using a program such as the 64-bit version of Scilab, which can read large data sets into volatile memory.

16.3.7 Update Models

We update the model when we wish to test something new or when we discover an error in the existing model. Simulations can be temperamental; for example, consider the climatic models used to assess the probability of anthropogenic (man-made) global warming (AGW). Some of these models are pessimistic, some optimistic, and some in between. We can only assess the quality of these models by assessing the quality of the assumptions driving the model.

16.3.8 Design Updates

Of course, any change to the product will require a change to the simulator that is simulating that product. We may see some reciprocation between the simulations and the design as the results of the simulation runs may drive design changes, which, in turn, drive further simulation runs. We recommend (again) that all versions of the simulator be under configuration management so that we can compare earlier designs with newer designs.

16.4 Objectives of Simulation

Simulations can be performed to analyze the behavior of a system. Not all simulations are used for automated testing. Regardless, general objectives consist of:

- Evaluation of various design concepts quickly without material investment
- Demonstration of system integration (peer reviews and customer feedback)
- Refinement of design concepts by predicting effects of performance parameters on system behavior
- Verification that the simulated object performs correctly across a wide range of nominal and fault scenarios
- Identification of key variables that impact the system and realization of the system implications, particularly with respect to potential interactions
- Reliability consequences
- Theoretical simulation results as reference for the practical verification

16.4.1 Evaluate Various Design Concepts Quickly without Material Investment

Using strong tools such as Matlab/Simulink or Scilab/Scicos, we can build our entire system virtually and simulate all significant functions simultaneously. If our model has been built well, we should get a fair representation of the operating product. We have seen this approach used with commercial and noncommercial vehicle design. This approach had the additional advantage of providing operating information to the suppliers. The suppliers, in turn, could bring their own models into the simulation—in essence, extending the customer simulation.

We can, for example, represent most of the mechanical and electrical features of a vehicle without any material investment other than the software, the understanding, and the computer on which to run the simulation. When we add in more specific simulations such as finite element analysis, SPICE (electronics), and computational fluid dynamics, we liven the simulation environment even further.

16.4.2 System Integration Demonstration

If we build a full-scale simulation environment using an advanced software tool like Matlab/Simulink, we can continue to use that tool when performing early system integration testing, where we may not have all the hardware for the new system. We can use the tool to drive what hardware we have and use the rest of it to simulate signals and other electrical and electronic data. The primary benefit is that we do not have to wait for a full set of hardware to begin our testing and designed experiments.

16.4.3 Refine Design Concept by Predicting Effects of Performance Parameters on System Behavior

Correct model development is critical to the success of our simulations, especially if we want to apply designed experimentation to the model or parts of the model. We can certainly do some element of parameter design using this approach. At the end of the parameter design simulation phase, we would expect to have recommended values for performance parameters that increase the robustness of the final product.

16.4.4 Verify That the Simulated Object Performs Correctly across a Wide Range of Nominal and Fault Scenarios

When we begin to run out tests on a simulated product, we will exercise *all* of the nominal possibilities and then proceed to potentially fault-generating scenarios. One of us did this kind of simulation and testing for the U.S. Department of Defense: We were controlling unmanned airborne and ground vehicles (drones). We wrote the simulator code before we ever wrote the real application code. The simulator code provided a virtual test bench against which we could test our real application code.

16.4.5 Identify Key Variables That Have an Impact on the System and Know the System Implications

We can use designed experimentation in the form of screening experiments to verify which variables are key. Some might object that we already know this from the controlling equations used in the simulation, but for a complex simulation, determining key variables in not that simple. Additionally, we can assess the importance of key factors (any factors, in fact) across the nominal range and outside the nominal range of values.

16.4.6 Reliability Implications

Because we can use our models to apply parameter and tolerance design before we have any hardware for testing, we can use the designed experiment method to improve the robustness of the product. We accomplish this goal by adjusting performance parameters into regions of behavior where they are relatively immune to noise and extraneous influences.

Depending on the sophistication of the simulation, we can also test our products under horrible circumstances we would not even consider executing in a full-featured laboratory. This ability provides yet another benefit to the simulation approach.

16.4.7 Theoretical Simulation Results as Reference for the Practical Verification

Once we have the real hardware, we find it prudent to execute confirmation runs to verify that our simulations do not have any serious misconceptions in the model. If we see significant differences between theory and reality, we either have a bad model or a product that may not have been fabricated correctly. Either way, we would then proceed to investigate for root cause and make corrections.

The more often we are able to generate realistic models for our final product, the easier it becomes, particularly if we created libraries of generic or commonly used components. This idea suggests that maintaining libraries for subsequent simulation efforts is beneficial to the design and testing efforts.

16.5 Simulation as Verification

Simulation offers some early verification possibilities. With the correct tools and the ability to model the component or system, it is possible to provide feedback on the design, making updates and alterations possible even while the product is in the early phases of development, as shown in Figure 16.9. One significant benefit of this approach can be the material and prototype part costs. Early in the product development life cycle, it becomes possible to critique the proposed design performance under the conditions and stresses allowed within the simulation environment. A review of this performance allows the design team to make adjustments to the product design even before money has been spent on the material. This capital includes prototype tooling costs and material and material handling costs.

16.5.1 Analysis of the Product Requirements

We recommend the use of the traceability matrix as a primary tool to analyze product requirements. If we use this tool for the creation of our simulations, we can basically check off which features our simulation incorporates, simultaneously indicating to ourselves what we are *not* doing and potential items for subsequent inclusion in the model. If we are truly fortunate, we may also detect items that should have been in the original specification or for which we need some kind of derivation.

16.5.2 Generation and Specification of the Simulation Concept

To specify our simulator, we need to know what it is we want out of the simulator; for example,

- Finite element analyses can be applied to heat or stress.
- Discrete event simulators do not have to run in real time.

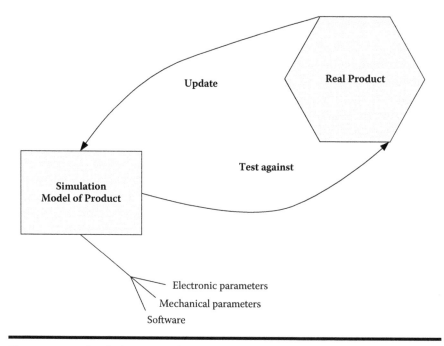

Figure 16.9   Simulation as verification.

- Actor-based simulators allow the actors to function relatively independently according to a handful of rules.
- Real-time simulators provide action much like what we would see exercising the real product, which makes these tools especially valuable for training.
- An estimation of the granularity of the simulated actions, coarse or fine.
- How much of the system we intend to simulate:
  - We can simulate entire systems; for example, commercial vehicle electronics (trucks).
  - We can build simulators for selected subsystems, which, if modular, can be used with a full system simulation.
  - Materials behavior (finite element analysis).

As with any other project, we will also do the following:

- Establish the goals for simulator.
- Create a simple statement of work.
- Schedule the development activities (if the simulator does not already exist).
- Collect all known vital parameters (nonvital if available).
- Establish criteria for evaluating the quality of the simulator.
- Define the future of the simulator (training tool, problem solving, etc.).

16.5.3 Assessment of the Simulation Concept

Our choice of tool will affect how we assess the capabilities of our simulator. For example, Matlab/Simulink and Scilab/Scicos are primarily oriented around discrete event simulation, although we see no reason why they cannot also be used for real-time simulation with the appropriate timing values. An actor-based simulator is going to go off and acquire somewhat of a life of its own. Of course, finite element analysis, computational fluid dynamics, and SPICE are all dedicated to specific functions relevant to their disciplines.

We may also want to see if our simulator can execute the same test cases we would use if we had complete hardware available. This approach is a powerful way to give ourselves an edge over the competition and accelerate test and evaluation. If we design our full-scale simulator appropriately, we can "hammer" the product night and day with thousands to millions of test cases, recording all results automatically for subsequent review. The payoff from a well-behaving simulator can be important to the future of the product and the future of business for the enterprise.

16.5.4 Generation of Test Scenarios

Test scenarios for the simulator will initially use nominal values so we have some level of confidence in the stated design range of the product. After that, our plan should allow for careful exploration outside the nominal range to establish the range of credibility for the simulator. As noted, we should have some criteria that allow us at least a crude estimation of the verisimilitude of the simulator so we can decide whether or not to progress with simulations.

If we are going for a substantial amount of verisimilitude, we can expect to be able to execute most of the test cases we would ordinarily run against the real product. Each time we conduct such a test, we are testing the virtual product and we are also testing the test suite. By the time we have the real product, the test suite should have matured and any bizarre issues should have already been removed.

16.5.5 Development of the Simulation Model

When building an initial version of a software-based simulator for the Department of Defense, we created a simple version of the final product that we called the *batch simulator*. Basically, this simplified version of the simulator would run with predetermined inputs. Although it had little flexibility, it allowed us to verify that our fundamental vehicle control code was functioning correctly (mostly in FORTRAN!). The display was also simple, using a Unix-based library called *curses* to create a text-based terminal on a personal computer. At this point, we had only invested about one to two weeks of work. We knew our basic control, guidance, and navigation code worked properly and we knew we could simulate basic turns and movement.

16.5.6 Making Available the Scenario Test Data and Performing the Simulation Runs

For the initial runs, we want to use nominal values so we can exercise the simulator in a region of relatively known expectations. We can then test the simulator outside the nominal values. When using values beyond the nominal range, we should verify that the model supports this range; otherwise, our results are likely to be meaningless.

Once we are reasonably satisfied as to the verisimilitude of the simulator, we can then proceed cautiously with our test scenarios. Any bizarre results from our virtual testing should be tried in the court of common sense. For example, the Department of Defense simulator behaved as our experience suggested it would behave. When trying to simulate the effect of wind or other environmental effects, we assessed our results with great care, as these effects were not understood nearly as well as the control, guidance, and navigation capabilities of the software.

16.5.7 Analysis and Evaluation of the Simulation Results

We should develop the criteria for evaluation of our simulator before we create the simulator. That way, even if we have no comparison product hardware/software, we still have some basis for accepting or rejecting the simulator. In the case of finite element analysis, computational fluid dynamics, and SPICE, we still want to compare our results from the simulation against expectations.

16.5.8 Testing the Simulation Model

We can verify that the simulator is operating; however, until we have product hardware or similar hardware against which to compare, verification of the simulator may be difficult. Nonetheless, we can continue to use the simulator as long as it does not violate any commonsense expectations.

16.5.9 Possible Upgrade or Modification of the Simulation Model

Any upgrade in the design of the product we are simulating requires an equivalent upgrade to the simulator if we intend to keep using the tool. The simulator should be under the same kinds of configuration management we are using for product hardware and software. We also upgrade or modify the model any time we discover an error in the model itself.

16.6 Simulation as Test Preparation

We can also use simulation as a tool for the preparation of all kinds of tests, including automated testing. This approach extends beyond requirements elicitation. The factors involved are the following:

■ Performance of the product itself
■ Performance of the test plan
■ Expansions of the test plan as we learn the product

16.6.1 Set Up Test Scenarios

Through simulation, we identify key performance attributes and variables. This, in turn, provides information we can use to perform physical and software testing. At a minimum, we should have some understanding of which factors seem to be most significant, and we can adjust our testing campaign accordingly.

16.6.2 Set Up Test Environment

The simulations may also provide some hints regarding the test environment. We may determine that some of the noise factors need more attention, and we make the attempt to convert a noise factor into a control factor.

16.6.3 Identify Key Test Measurable and Instrumentation Needs

We typically use a spreadsheet to specify the measurable values and the instrumentation required. With a little bit of extra effort, we add the *next* calibration dates. We also add the measurement uncertainty value if we have that information.

We also use the spreadsheet format to clearly specify test requirements such as

■ Rise times
■ Maximum values
■ Dwell times
■ Minimum values
■ Falloff times
■ Number of cycles

In some situations, we find it makes sense to also clearly state the *purpose* of the test. For example, we have had many customers require a test called a *swept sine test*. With this test, we cycle a sample through a variety of frequencies and displacements on a vibration table. Many customers believe this is a test to failure. We use it as a means to determine the resonant frequencies of the product so we can use that information later for accelerated testing.

16.6.4 Human Resource Needs

Often, the complexity of the kind of testing we are doing determines the kind of human support we need to accomplish our mission. In many test facilities, technicians are adequate to the requirements. However, in our facilities, we use a higher proportion of test engineers because we often need to develop software and fixtures to test the part as well as perform part of the analysis.

16.6.5 Material Resource Needs

Just as we need human support, we also need sufficient material to perform our tests. Examples of this kind of material are the following:

- Test material (e.g., Arizona fine dust for the dust tester)
- Lubricant to maintain the machines
- Refrigerants for thermal chambers
- Various salts for salt spray and fog
- Pressurized air
- Pressurized water
- Hazardous material containment items
- Fire extinguisher
- Dehumidifier/humidifier to control humidity levels to standard
- Timers
- Recording material for videos of test
- Cameras
- Sample material (units under test)

16.6.6 Test Sequencing

If we are using a simulator as an advance preparation tool for the hardware testing, we should sequence our virtual testing in the same way we would if we were running the actual product through testing. If our model is extremely sophisticated, we may even be able to simulate some level of degradation of the product as we move through the test sequence. This situation is particularly true if we are able to use known physical models and derive the appropriate equations to define the level of degradation; by adding a random factor, we can increase the verisimilitude of the simulation and provide ourselves with some food for thought before we ever test the hardware.

16.6.7 Identification of "Passing" Criteria

We should be using the same criteria for assessing failure for a simulation as we would for the actual hardware or software. If we do not use the same criteria, our simulation may not be providing close enough replication of the actual product.

16.7 Conflict between Simulation and Test Results

Because simulators are actually instantiations of models, we will occasionally see a conflict between the abstraction of the model and the reality of actual product testing. On the testing side, we would review our test assets, our measurement methods, our tactics, and the operational environment. On the simulator side, we would review the model for accuracy and identify any missed parameters as well as the range of the previously identified parameters.

16.7.1 Emulation

Emulation is not the same as simulation. With emulation we are using software and/or hardware to try to provide a more or less *exact* replica of the device under test. We discuss in-circuit emulators elsewhere in this book. We have also used software-only emulators that emulate a specific microprocessor on a personal computer. This approach is somewhat easier with eight-bit processors and becomes more difficult as the size of the processor increases.

To some extent, a byte-code compiler provides a pseudo-processor usable on any architecture for which it has been developed. The Java language, among others, uses this approach to proliferate itself onto new architectures. The Java Virtual Machine (JVM) is the pseudo-processor for each Java implementation.

Chapter 17

Span of Tests

17.1 Software

When testing software, we will follow the pattern exhibited in Figure 17.1, working our way from the bottom to the top. We also shift from the developers themselves running the unit tests to an internal verification and validation team, to the customer testing the final product.

17.1.1 Unit Test

The lowest portion of the software is a unit. This is where the testing should start and is often performed by the software developer. This testing of units occurs before the build of the various components that make up a revision of software. The developer will pass valid and invalid (or unexpected) inputs to the unit to make sure that it performs as expected. Nothing should be unexpected or unaccounted for within this lowest level of software. In other words, the unexpected inputs do not generate a catastrophic response from the unit.

17.1.1.1 Build Tests

A typical software build test will exercise only the functions and subroutines that have been changed in the most recent build—testing the rest of the code occurs during regression testing. Basically, we conduct a quick build test to verify that our changes seem to function according to design and that no anomalies occur. As with most testing that displays no failure modes, a build test without failures can lead to a false sense of confidence about the quality of the software.

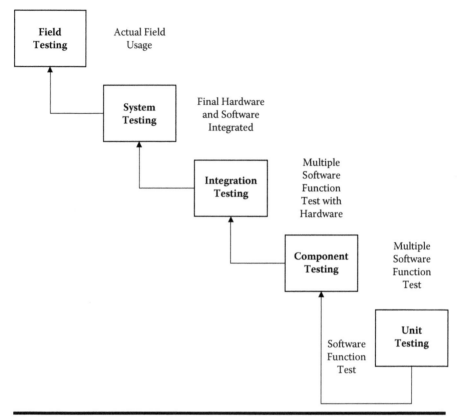

Figure 17.1 Software test hierarchy.

17.1.1.2 Regression Testing

We perform regression testing at the end of each software build in order to verify that we have not adversely affected any of the previously existing code in our product. Incorrectly architected software can exhibit "side effects"; that is, unexpected behavior that occurs after a change because functions and variables are too closely coupled. A well-designed suite of regression tests will usually be automated whenever possible. The test engineer executes a "once-over" that the regression tests provide and then reports the results. The sad side of regression testing is that a regression test that evinces no failures is also a suite of tests that tells us nothing and can provide a false sense of confidence regarding the stability of the code.

17.2 Component Test

With component testing, we have now fused some of our units into a *component*, which is basically whatever we define it to be but is definitely not a unit! What we are doing is traversing our tree to higher and higher levels of complexity, assuming that we are practicing some modicum of bottom-up testing.

17.3 Subsystem-Level Integration Testing

By the time we begin integration testing, we should have software that resembles somewhat of a product and hardware on which to test. This is the point where we begin to see software and hardware interactions. Integration testing is a critical milestone in our development.

17.4 System

At the system level, we usually have multiple hardware components, data buses, and a multitude of opportunities for interaction. We are most likely to see serious timing misbehavior during system testing because each hardware subsystem or component usually has its own clock. In some cases, we have seen bus speeds moderately relaxed to make the system work.

17.4.1 Subsystem Test

The word *system* can be confusing. Are we speaking of the final system or the software as a system? When we discuss a subsystem and software in the same breath, we mean that the software is being treated as if it were a complete subsystem in itself. The merging of hardware and software leads to the integration test followed by the system test. All of these terms should be defined:

1. First, define by the development team.
2. Second, include the suppliers and management.
3. Third, secure agreement from the customer.

17.4.2 System-Level Integration Test

Figure 17.2 shows a graphical representation of one common approach to software testing. Basically, we move down the left side as we go through our hypothetical design process, which culminates in "code." Once we have code, we begin testing on

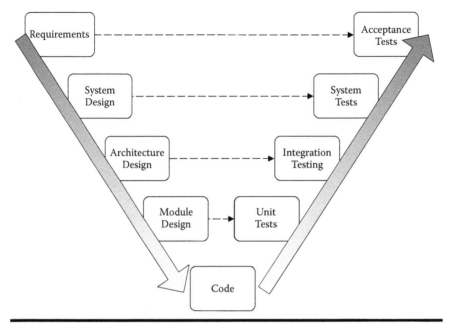

Figure 17.2 Standard testing V-model.

the right side, moving from small component tests up the scale until, finally, we are performing full system testing. Of course, this model is extremely idealistic. Generally, software development will get caught in a loop of test, analyze, and fix (TAAF) as we begin to flush out defects.

Figure 17.3 shows another graphical representation of an agile development approach to software testing (a la Scrum). This method relies on quick releases with frequent testing of the product and evaluations from the customer. This approach downplays the specification demands and builds on frequent communication between the team members in addition to significant testing of each iteration or increment of the product.

Each approach can achieve the goal of delivering a quality product. To make a sports analogy, if a football team has a strong running game, that will be the approach the team would likely take to success. If the team is better equipped to pass, and has a limited running game, it would not make much sense to use this limited running game. Selection of method is based on the strengths and weaknesses of the organization developing the product and the level of sophistication of the product being developed.

17.4.3 System "Live Test"

Sometimes live testing includes only hardware-in-the-loop testing; at other times, the software and the product in which it resides will be installed on the final product

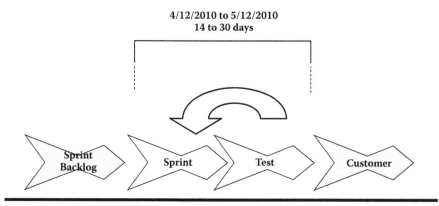

Figure 17.3 Iterative development model.

(e.g., a motor vehicle). One of the most significant difficulties with this kind of testing is ensuring that the system sees enough variety during the test interval. Some commercial vehicle manufacturers will drive special trucks around the country for intervals from 90 days up to a year to see if the various subsystems behave appropriately. Achieving a meaningful melange of environmental challenges is a daunting task.

17.4.4 Other System Tests

We have a few other options for system-level testing:

■ *Factory acceptance testing* (FAT) occurs when a customer decides to run a formal set of tests to verify the product functions as *they* expect it to function.
■ *Operational testing* (OT) is a term used by the Department of Defense and it is roughly equivalent to "live test."
■ *Manufacturing qualification testing* (MQT) occurs when we have means for showing that the product has journeyed through production with no anomalies.

17.5 Production Test

Production testing and MQT are often synonymous. The purpose is to verify that the manufacturing lines are running correctly by assessing the correctness of the product. The most common difficulty occurs with small sample sizes, especially if the product is not sellable after testing.

Another form of production test is the occasional ongoing sampling to check for manufacturing line drift. While uncommon with software, it is routine with hardware.

17.6 Static Analysis

The Unix tool *lint* is a typical example of a static analysis application. When run on existing code, it will flag well-known errors and potential errors for further investigation. The most sophisticated version of this kind of tool can help convert the mundane C language into a highly typed language, much like the Department of Defense language, Ada.

17.7 Structural Analysis

We can do structural analysis on software by looking at the call tree for the various functions and subroutines. If no free or commercial tool is available, we can quite easily write our own code to pull out the function names, called *functions*, and use a graphical tool like AT&T's Graphviz to represent the tree graphically. Some compilers will generate a call tree for a log file, making the home-grown software even easier to implement.

Some integrated development environments know enough about specific languages that they can perform an action called *refactoring*. When refactoring, we change a program's source code without altering its external functional behavior so as to improve some of the nonfunctional attributes of the software, such as readability.

17.8 Simulation

We can use computational models to provide advanced simulation of certain properties of our product. For example,

- Computational fluid dynamics can model convection and liquid flows.
- Finite element analyses allow us to stress a part virtually in two dimensions—temperature and mechanical stress.
- SPICE allows us to model electrical circuits.
- Discrete event modeling with tools like MATLAB®/Simulink allow for overall system modeling.
- Actor-based modeling with tools such as NetLogo also provide system modeling capability.

Profound and knowledgeable use of these tools can lead to a reduction in hardware test times. Furthermore, the designers can run virtual designed experiments on their models.

17.9 Prototyping and Simulation

Prototypes are something we build in software or hardware. A software prototype is a working software system that may not see the light of day as a complete product. A hardware prototype should be at least partially if not wholly functional—it also may not become a full-fledged product. An alternative to prototyping is to use simulation, usually with some kind of software tool. We do not want to confuse simulation using software with software simulation for the purpose of showing the product functioning to a customer.

Prototypes have the benefit of being testable in much the same way as the real product would be tested. A prototype is not the same thing as a model. For example, when we use stereographic lithography to build resin-based models, they are typically useless for any kind of mechanical testing because they are not robust enough. As a model, however, they can provide an idea of the fit and form of the potential product. Prototypes can become quite expensive if we are using mechanical components because we will most likely have to use what is known as a *soft tool* to build them. A soft tool can mold parts that are very high quality—the soft tools simply do not last as long as a hard tool.

Simulations have the benefit of not needing a soft tool to build the parts; we can build simulations virtually on a computer. We may have to use special software if we are trying to model electrical/electronic circuits or mechanical parts. Typical examples of such software are tools that support SPICE (electronic), finite element analyses, and computational fluid dynamics. With a software simulation, we can tweak various parameters almost at will. Furthermore, we can also run designed experiments against our simulated parts and see what happens as we vary the experimental factors.

The use of a simulation allows us to test a product to destruction innumerable times while only incurring the cost of computer time and programmer effort. Even soft tools can run easily into the thousands of dollars. If we have an existing simulation test bench, we can save those thousands immediately, not to mention the benefits of doing hundreds of test runs using software alone. With a good simulation, we should have no excuse for not doing our designed experiments on the virtual products. Figure 17.4 shows how we can use two "threads" of practice to produce the final product. The benefit, of course, with simulation is that we can beat up the product without consuming potentially expensive material. We are able to conduct experiments that might be unsafe or impossible with the real material.

17.10 Reliability and Confidence

Reliability and confidence in software statistics can be problematic due to the paucity of usable models. Only the Rayleigh distribution approach appears to have much support in the literature. We have seen this approach used in the early 1990s (the

Hardware thread

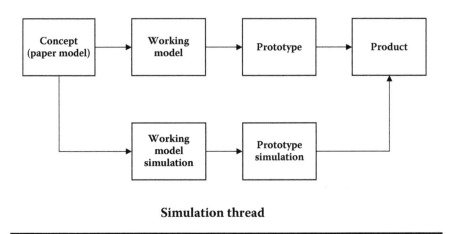

Simulation thread

Figure 17.4   Prototype and simulations.

customer did not know what they were looking at, but the probability distribution function was obvious and beautiful!), and we have used it to model the status of the software across multiple builds. Another issue that can cause problems with this approach is whether we are doing full validations or TAAF-style validations (where we do not perform a complete assessment of the software). Combining the two modalities can produce plots that tell us very little and distort the reality of the quality of the software. A topic for research would be to see if we could normalize the defect quantity against the number of test cases and then determine if the results even make sense.

17.11 Limitations of Reliability and Confidence

17.11.1 "Normal Distribution" Assumption

When we perform reliability testing and hypothesize some supposed distribution (usually a continuous distribution), we almost never see results that resemble a normal distribution. In the rare case when we do see such results, we generally theorize that the product is in a rapid wear-out scenario.

One of the most common continuous distributions used in reliability analysis is the Weibull distribution, a three-parameter distribution that can assume a variety of shapes and locations, making it an extremely flexible tool for the reliability engineer. Figure 17.5 (output from Minitab) shows a few of the shapes that a Weibull distribution can assume.

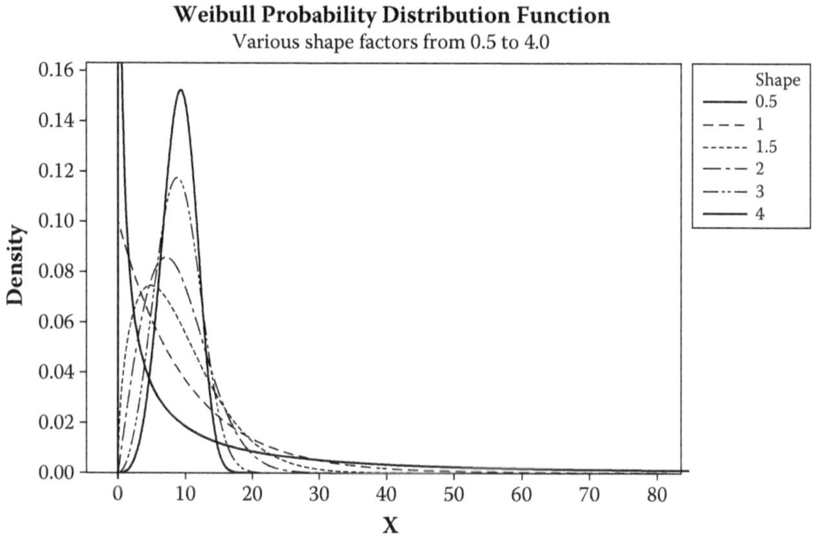

Weibull Probability Distribution Function

Various shape factors from 0.5 to 4.0

Figure 17.5 Weibull distribution appearance.

17.12 Concept of "Life" and the Product

The life of the product is the expected duration the product is to perform the functions. The traditional approach to representing the life of the product is the so-called bathtub curve, which represents a theoretical consideration of electronic failures (see Figure 17.6). The product starts out with some level of infant mortality, which declines quickly and is followed by a period of random failures, which, in turn, is followed by an increasing failure rate as the product ages. Modern electronic products do not follow this curve very well, but it does provide a theoretical construct for comparison. We have seen components fail with the first two modes; for example, "pogo pins" on test devices have springs that hold them against the test points. The failure scenario for pogo pins had a very high level of infant mortality followed by a long period of random failures. These particular results made it difficult to instantiate a preventive maintenance plan other than daily inspection of the test devices. Figure 17.7 shows the distribution derived from empirical data on these parts.

Figure 17.7 is roughly exponential in appearance, which is typical of components that fail randomly. We found that an aggressive pogo pin replacement program asserted some modicum of control. Once we had this level of control, we switched to a statistical analysis of *which* pogo pins were failing and created a spares bank based on that data.

Please note that software does not assume a bathtub curve appearance because it does not wear out. The difficulty with software is that all copies of a guilty version will have the same bad properties, which is why defective software can produce massive product recalls.

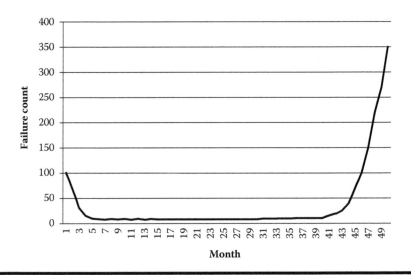

Figure 17.6 Product life.

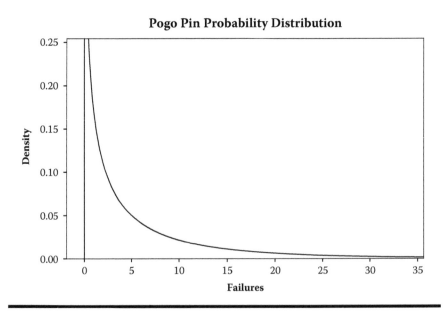

Figure 17.7 Product life analysis for pogo pins.

17.13 Establishing Product Life Exposure

Knowing how long the product should function (live) makes it possible to determine the full range of stimuli the product will be subjected to during this life. These will become the requirements the product must be able to meet. In some organizations, there are groups that capture this product requirement information; in others, the verification group will be part of requirements elicitation.

In the automotive world, original equipment manufacturers (OEMs) have test tracks that stress the vehicle and the components. Specific failures with similar products (in similar locations on the vehicle) in the field can be correlated to the number of laps or amount of exposure on the track. For example, say a bracket fractures after 100,000 miles in the field. The same bracket fractures similarly after 25 cycles on the track. That suggests that each cycle on the track corresponds to 4,000 miles of real-world stimulus for that bracket in that location. Keep in mind that we will need to test enough parts to apply a level of confidence to our correlation and, as always, we need to remember the limitations of correlational approaches.

17.13.1 Requirements

We can install instrumentation to measure the environments in which the product will perform its function and record the information about the environment as we operate the product. Figure 17.8 shows the gathering of information on a specific stimulus for a unit of time. The exposure during this time can be extrapolated cautiously to get a sense of the total exposure during the life of the product.

Any transient information needed must be considered when collecting data in this way. For example, seasonal and geographical changes may have an impact on thermal and moisture stimuli. In the automotive world, the requirements for a specific product may be determined by installing instruments on a number of vehicles with a variety of applications over a geographic distribution to gather this information. Some examples would be

- Amount of time windshield wipers are used and type of use (intermittent) per unit time
- Specific ABS events in cold climates
- Specific function activation (such as HVAC blower) per unit time

17.13.2 Measuring the Outside World

To really understand the outside world implications on the product, it is necessary to understand the outside world behaviors. We do this by finding the boundary conditions to which we believe the product will be exposed. For example, to understand the highest temperature extreme would require going to a place where you both believe that the product will be used and that has an elevated temperature. We use

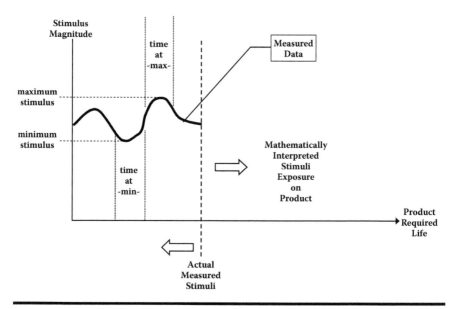

Figure 17.8 Life exposure requirements.

instrumentation around the area where the product will be used to record various stimuli. This approach is not only valid for thermal constraints or impacts but also for vibration, ultraviolet damage, dust storms, and any other real-world stimulus that makes sense.

Until we know what the product will experience, it is really just guesswork whether our quality will meet what is expected by commercial customers and end users. If we are a supplier, the customer may tell us what the expected product environment will be; however, we have experienced many cases where this information can be dangerously inaccurate. We know about one example where a customer never told the supplier that the product would be installed on the "rail" of a commercial vehicle, exposed to the elements, especially road salt. The sampling population can get quite high to ensure meaningful representation of the field stimuli.

In some cases, the customer may have developed their own standards or we may have industry-specific standards that describe the environment where a product will be used. We think it appropriate to incorporate severe environments into a design failure mode and effects analysis in situations where we are certain the product will be exercised in less than benign settings.

17.13.3 Standards

The use of standards in testing a double-edged sword. Blindly using standards can cause more harm than good. This is true when the standard does not really reflect the customer's environment or use of the product. Specifications have similar issues with testing to requirements. If the standards were not generated out of real-world information from a sufficient population of parts, it is likely that the product will experience some field issues unless we are lucky. We do not consider relying on luck to be a prudent approach to product development.

On the positive side of standards, we find a set level of performance for the product or the test protocol. Application of industry standards means that the product will perform to some industry standard, which has a homogenizing effect on the product quality from supplier to supplier. In other words, they all know the rules of the game and generally abide by these. If we begin to apply substantial amounts of custom testing, we may end up with custom parts, which can become expensive.

Clearly, the use of standards requires understanding the following:

- Our product
- Any product on which our product will be installed (if any)
- Customer needs (not just specifications or requirements)
- The boundaries of the customer's expected environmental use
- An estimate of potential abuse the product may see

We want to balance the regularity of standardized testing with our desire for complete product characterization while simultaneously living within our budget.

Chapter 18

Exit Criteria

We use exit criteria to ensure that we have some form of closure to our testing strategy for a specific product. This product or service may receive further testing but, if we are part of a development sequence, we must have well-defined criteria that tell us when the product is sufficiently well developed to be released to customers.

If we are using the failure mode and effects analysis (FMEA) tool during product development, we will ascertain if we have developed our verification and validation plans to respond to this document. Because the FMEA approach is a systematic effort to eliminate significant issues before they actually become problems, we need to ensure that our test documents are designed to reflect this need.

18.1 When Is Enough, Enough?

If we do not define "enough," we can test forever, trying to achieve

- Adequate sample sizes
- Sufficient durations to reveal time-based flaws
- Sufficiently random representation of the population of parts

No matter how important these considerations may be, we must choose a point at which we make a business decision to release the product. We might support our decision, as with software, by comparing our defect arrival rates with a Rayleigh distribution—releasing when we are well out in the tail under the assumption that this model is telling us that our defects are few.

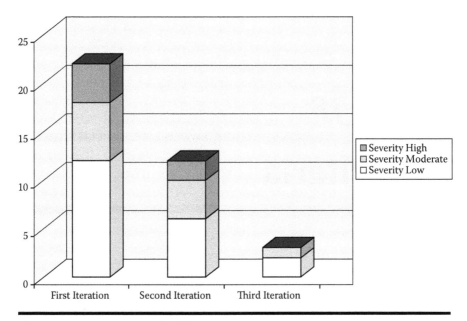

Figure 18.1 Testing completed?

18.2 Compared to Plan?

One of the reasons we plan is to provide a means for controlling the scope of our testing. Even if our testing is less than perfect—testing is always less than perfect—we still have recourse to our plan to control the sequence and completion of activities. A plan does not mean that we are locked into a specific course of action, but it provides structure as well as a benchmark against which we can compare our actual set of actions. We can also build into our plan the increase in the number of test cases and the concurrent decrease in remaining issues found (see Figure 18.1).

We say that testing is never perfect because we do not have the following needs ever satisfied:

- Monetary resources are limited.
- Human resources are often scheduled for other activities.
- Important documents like the design (DFMEA) may not be constructed well and updated promptly.
- Sample sizes are usually too small.
- Statistical confidence is often a nearly meaningless value.

We are not saying that we should not test. What we are saying is that we must understand the limitations of testing. We must ensure that we are using all the tools at hand to produce a great product.

18.3 Testing Ethics

18.3.1 Test Plan Ethics

The minimum test plan we must execute against a product will be derived from the customer specification or our own standard, if we have one. We also recommend that our test group conduct product tests to failure and thence to destruction (if destruction makes sense in the characterization of the product). We should submit the compliance test plan for approval by the customer. This document may also contain some tests, verifications, and validations we have recommended.

The characterization test plan describes activities we do for our own benefit. It is not necessary that this document be revealed to the customer. They did not ask for it, they are not paying for it, and they should not get it. This test plan is our bulwark against internal and external design and manufacture ineptitude.

18.3.2 Test Report Ethics

When we execute a test plan, we generally update the plan in such a way that it transforms into the report—this way, all the relevant information will be contained in a single document. For any test required by the customer, we will indicate the results honestly and clearly. In some ways, we are nicely positioned in the enterprise because we report what is wrong—it is not our responsibility to fix the issues we discover. On the other hand, we have a tremendous responsibility to report everything we see.

The issue becomes more interesting when we execute our internal test plan. Now we wish to report the results as close to concurrently with the testing as we can so that the design people are receiving updates about the product in nearly real-time. For our design verification testing to have any meaning, we must execute meaningful tests as soon as we have working models, prototypes, or parts. For example, we do not have to wait for tooled mechanical parts to commence execution of our characterization testing because we can still test the electronics.

18.3.3 Customer Relationship Ethics

18.3.3.1 Communication of Faults

18.3.3.1.1 Before Launch

Before the launch of a new product, our team should communicate discovered faults only if they need customer input or if the correction of those faults will cause a launch delay. Otherwise, we see nothing unethical in not reporting faults that will be corrected unless we have people using prototype parts who might become injured from the fault.

Prototypes and working models should be clearly identified as such before they leave our plants. If the customer does not want these pieces to be identified this way, we need a written release from the customer before we can ship prelaunch parts to them. At this stage, when we are shipping prelaunch parts, we should indicate

- Current status of the development
- Current status of the testing
- Potential field issues with unimplemented features
- Potential field issues with features implemented incorrectly

18.3.3.1.2 After Launch

If we detect a fault after launch, we must report it to the customer. We are now selling the product, and it would be unconscionable to do otherwise, particularly if we are dealing with any safety issues. In some cases, our reports on defects or faults may initiate a campaign for or a recall of the product. If the customer finds the fault significant enough for the black-eye of a recall, we can consider it significant. Although recalls are financial catastrophes, they are far, far better than the lawsuits that will follow tragic accidents.

We find it unsalutary to conceal negative test results from customers after the product has launched and we have material in the field. While we have sometimes seen our customers become extremely unhappy, in the final analysis, the entire incident proceeds more smoothly than might otherwise have been the case. In situations where we have seen executives attempt to "spin" the topic, the ultimate customer results were consistently negative. Do not even try to spin a product fault!

18.3.3.2 Launch Delays

Any test issue that will potentially result in a delay of the product launch must be communicated to the customer at the earliest possible moment. The customer often has their own set of launch plans, and a delay in launching our product may throw off their schedule. Once we have reported the issue, the situation then becomes a commercial problem. If we are working for a large enough enterprise, we will have a marketing department that is more competent than we are to deal with the by-products of the launch delay.

Although launch delays can result in customer dissatisfaction, we feel that it is better to run the product through a complete test suite rather than try to abbreviate the test sequences to appease the marketing staff. If our documentation shows that we slacked off or that we rushed, we may have problems if we ever have to go to court. Hence, we suggest that our organization negotiate a launch delay with the customer that allows for completion of all testing up through system testing.

18.4 Final Words

We feel as if we could have written many more words than are contained in this book; however, we already know that testing is infinite. Nobody ever truly completes testing—they just run out of time. We want to make sure that we at least achieve a responsible amount of testing that we can define with the following:

- Use of FMEA
- Product characterization
- Test to failure
- Test to destruction
- Instantiation of relatively *realistic* tests

In all cases, our reports should be brutally honest, reporting every anomaly and observation that we see. Obscuring test data and conclusions is unethical and could lead to severe repercussions if our organization has to go to court over product misbehavior.

We, as test engineers, are the gatekeepers. It is we who raise our metaphorical lances and say, "You shall not pass through here." Our responsibility is always to

- Protect our customers
- Never send trash to a customer (internal or external)
- Practice safety first!

Bibliography

Alexander, Christopher. *The Timeless Way of Building* (Oxford, UK: Oxford University Press, 1979).

Deleuze, G. and F. Guatarri. *A Thousand Plateaus: Capitalism and Schizophrenia* (Minneapolis: University of Minnesota Press, 1987).

French, Simon. "Cynefin: Repeatability, Science and Values," in *Newsletter of European Working Group, "Multiple Criteria Decision Aiding,"* Series 3(No. 17), 1, Spring 2008.

Gawande, Atul. *The Checklist Manifesto: How to Get Things Right* (New York: Metropolitan Books, 2009).

Krishnamurti, Jiddu. *Dissolution Speech.* Available at: http://www.jkrishnamurti.org/about-krishnamurti/dissolution-speech.php (accessed May 1, 2010).

Nietzsche, Friedrich. *The Will to Power,* translated by Walter Kaurmann and R. J. Hollingdale (New York: Random House, 1967).

Nietzsche, Friedrich. *Twilight of the Idols,* translated by R. J. Hollingdale (London, UK: Penguin Books, 1968).

Sackett, David L. "Bias in Analytic Research," *Journal of Chronic Diseases* 32(1-2), 51–63, 1979.

Index

Milton Keynes UK
Ingram Content Group UK Ltd.
UKHW031128141024
449569UK00006B/353